Bernard Kipsang Rop
Wycliffe Habel Namwiba
Naomi Chepkemoi Sang

Sedimentary Rocks

Bernard Kipsang Rop
Wycliffe Habel Namwiba
Naomi Chepkemoi Sang

Sedimentary Rocks

A Comparative Illustrative Approach

LAP LAMBERT Academic Publishing

Imprint
Any brand names and product names mentioned in this book are subject to trademark, brand or patent protection and are trademarks or registered trademarks of their respective holders. The use of brand names, product names, common names, trade names, product descriptions etc. even without a particular marking in this work is in no way to be construed to mean that such names may be regarded as unrestricted in respect of trademark and brand protection legislation and could thus be used by anyone.

Cover image: www.ingimage.com

Publisher:
LAP LAMBERT Academic Publishing
is a trademark of
Dodo Books Indian Ocean Ltd. and OmniScriptum S.R.L publishing group

120 High Road, East Finchley, London, N2 9ED, United Kingdom
Str. Armeneasca 28/1, office 1, Chisinau MD-2012, Republic of Moldova, Europe
Managing Directors: Ieva Konstantinova, Victoria Ursu
info@omniscriptum.com

Printed at: see last page
ISBN: 978-620-8-44043-5

TABLE OF CONTENTS

PREFACE 3

ACKNOWLEDGEMENTS 5

INTRODUCTION 7

LEARNING OUTCOMES AND SPECIFIC COMPETENCIES 7

CHAPTER ONE. WEATHERING PROCESSES OF SEDIMENTARY ROCKS 13

1.1. Introduction 13

1.2 Physical Weathering 14

1.3 Chemical Weathering 16

1.4 Biological Weathering 34

1.5 Factors Affecting Weathering Processes 44

CHAPTER TWO. TRANSPORTATION, EROSION AND DIAGENESIS 49

2.1. Running Water 49

2.2. Moving Ice (Glaciers) 50

CHAPTER THREE. CLASSIFICATION OF SEDIMENTARY ROCKS 56

CHAPTER FOUR. CHARACTERISTICS OF SILICICLASTIC ROCKS 62

CHAPTER FIVE. CHARACTERISTICS FOR NON-SILICICLASTIC ROCKS 69

CHAPTER SIX. MAJOR PHYSICAL PROPERTIES OF SEDIMENTARY ROCKS 75

CHAPTER SEVEN. SEDIMENTARY STRUCTURES 81

CHAPTER EIGHT. COMPARATIVE ILLUSTRATIONS 87

8.1 Siliciclastic Rocks 87

8.2 Non-Siliciclastic Rocks 93

CHAPTER TEN. EXERCISES FOR SENDIMENTARY ROCKS 97

REFERENCES 123

ABOUT THE BOOK 128

PREFACE

The study of sedimentary rocks is essential for understanding the Earth's processes, resources, and history. *Sedimentary Rocks: A Comparative Illustrative Approach* serves as a comprehensive resource for students, educators, and practitioners in geology, particularly for learners at the Certificate and Diploma levels. This book also aims to support Artisanal Small-scale Miners (ASM) by providing essential knowledge to enhance their skills and understanding of geological aspects relevant to their work.

In this book, we explore fundamental concepts of sedimentary rocks, including their formation, classification, and the processes that shape them. Each chapter builds upon the previous one, guiding readers through the complexities of weathering, erosion, transportation, and diagenesis. We present the material clearly and engagingly, utilizing a competency-based approach aligned with UNESCO's ASM Certificate and Diploma Programmes.

A key feature of this book is the integration of comparative illustrations, which juxtapose schematic drawings with real-world field images. This visual approach enhances comprehension and bridges the gap between theoretical knowledge and practical application, facilitating a deeper understanding of sedimentary processes and their implications.

To enrich the learning experience, we have included a variety of questions and exercises throughout the text. These are designed to encourage critical thinking and allow readers to assess their understanding and engage actively with the material.

We acknowledge the contributions of many individuals in the development of this book. The insights and research of numerous authors have been invaluable in shaping our discussions and illustrations. We extend our heartfelt gratitude to these contributors, as well as to our colleagues and friends who have supported us throughout this journey, especially during the challenges posed by the Covid-19 pandemic.

As you explore the world of sedimentary rocks, we hope this book serves as a valuable resource that enhances your knowledge and inspires a greater appreciation for the dynamic processes that shape our planet. Sedimentary rocks are not merely geological formations; they are archives of Earth's history, revealing the intricate interplay of environmental conditions, biological evolution, and geological events over millions of years.

Whether you are a student embarking on your academic journey, an educator seeking to enrich your curriculum, or a practitioner in the mining and construction industry, we believe the insights contained within these pages will be beneficial in your endeavors. The comprehensive analysis and illustrative comparisons presented here aim to equip you with a deeper understanding of sedimentary processes, the formation of various rock types, and their significance in both natural and human contexts.

Moreover, the knowledge gained from this book can empower you to make informed decisions in your professional practice, whether it involves resource extraction, environmental management, or educational outreach. Understanding the properties and behaviors of sedimentary rocks is crucial for sustainable practices in mining and construction, where responsible stewardship of resources is paramount.

Finally, we invite you to delve into the fascinating dynamic systems and processes that sedimentary rocks reveal about the Earth's history and the resources they provide for our future. As you engage with the material, consider the broader implications of your learning. Reflect on how sedimentary geology intersects with contemporary issues such as climate change, land use, and resource sustainability. The stories embedded within these rocks are not just relics of the past; they are vital narratives that inform our present and guide our future.

In your exploration, may you find inspiration in the complexity and beauty of sedimentary geology, and may it ignite a passion for discovery that drives you to further investigate the natural world. Together, let us celebrate the rich tapestry of Earth's geological history and work towards a sustainable future that honors the resources and knowledge we inherit from our planet.

Geol. Prof. Rop Bernard Kipsang, PhD

Reg. Geologist No.1 Kenya

ACKNOWLEDGEMENTS

This book, *Sedimentary Rocks: A Comparative Illustrative Approach*, aims to provide a comprehensive overview of the major minerals and rocks commonly encountered in quarrying and mining sites. It is specifically designed to meet the educational needs and interests of learners at the Certificate and Diploma levels in the field of Geology. Additionally, this book serves as a valuable resource for Artisanal Small-scale Miners (ASMs) who are looking to enhance their knowledge through appraisal short courses, equipping them with essential insights into the geological aspects of their work.

The structure of the book is organized into several key chapters, each focusing on essential aspects of sedimentary rocks. The chapters covered include: Learning Outcomes and Specific Competencies, Weathering Processes of Sedimentary Rocks, Transportation, Erosion and Diagenesis, Classification of Sedimentary Rocks, Characteristics of Siliciclastic and Non-Siliciclastic Rocks, Major Physical Properties, Sedimentary Structures, and finally, Comparative Illustrations of Rock Types, Occurrences, and Uses. This comprehensive organization ensures that readers can systematically build their understanding of sedimentary geology.

In developing this book, the authors adopted a competency-based and evaluative approach to present the material effectively. The expected learning outcomes and specific competencies outlined in the book are aligned with the standards set forth in UNESCO's ASM Certificate and Diploma Programmes, ensuring that the content is relevant and applicable to both academic and practical contexts. This alignment not only enhances the educational value of the book but also ensures that learners are equipped with the skills necessary for success in the field.

A significant effort was made to bridge the gap between field realities and classroom learning. This is exemplified in the comparative illustrations provided throughout the text, which feature schematic drawings alongside field images. These visual aids are designed to enhance understanding and facilitate the application of concepts learned to real-world situations, making the material accessible and engaging for learners at all levels. By integrating visual elements with textual explanations, the authors aim to cater to diverse learning styles and promote a deeper comprehension of sedimentary processes.

The authors have also included a variety of questions aimed at evaluating the reader's comprehension and ability to apply the concepts discussed. These questions emphasize practical applications and encourage learners to think critically about the material, fostering a deeper understanding of sedimentary processes and their implications in mining and geology. This interactive approach not only reinforces learning but also prepares readers to tackle real-world challenges in their respective fields.

We would like to express our sincere gratitude to the numerous authors whose vital knowledge and illustrations have been referenced throughout this book. Their contributions have significantly enriched our work, and we acknowledge their professional efforts in advancing the field of geology. In particular, we extend our appreciation to the authors cited in the references at the end of this book, whose research and insights have provided a solid foundation for our discussions and illustrations.

Special thanks are also extended to our colleagues and friends for their unwavering support and encouragement during the challenging times brought about by the Covid-19 pandemic. Their belief in this project has been invaluable, and their insights have helped shape the content of this book. The collaborative spirit and shared commitment to advancing geological education have been instrumental in bringing this work to fruition.

Finally, we give glory to the Almighty God for providing us with the energy, motivation, and resilience needed to overcome the challenges posed by the pandemic. It is through divine guidance that we have been able to bring this print medium to fruition, and we are grateful for the opportunity to share this knowledge with practitioners in the mining and construction industry, as well as with aspiring geologists and artisanal small-scale miners who seek to enhance their skills and understanding of sedimentary rocks. This book is not just a compilation of knowledge; it is a testament to the collective effort and dedication of all those involved in its creation, aimed at fostering a deeper appreciation for the geological sciences and their practical applications.

Prof. Rop B. K., Mr. Namwiba W. H. & Ms. Naomi C. Sang

INTRODUCTION

LEARNING OUTCOMES AND SPECIFIC COMPETENCIES

Expected Learning Outcomes

By the end of this book on sedimentary rocks, readers will have developed a comprehensive understanding of the various aspects of sedimentary geology. The following learning outcomes outline the key competencies and knowledge areas that will be covered, enabling readers to engage with the subject matter in a meaningful and informed manner.

1. Understand Natural Processes

Readers will demonstrate an understanding of the natural processes and agents that contribute to the weakening of rock masses. This includes:

- Physical Weathering: Understanding the mechanical processes that break down rocks without altering their chemical composition. Key processes include frost wedging, thermal expansion, and exfoliation.
- Chemical Weathering: Recognizing the chemical reactions that lead to the alteration of minerals within rocks, including hydrolysis, oxidation, and carbonation. Readers will learn how these processes contribute to the formation of secondary minerals and the overall stability of rock masses.
- Biological Weathering: Exploring the role of living organisms, such as plants, animals, and microorganisms, in the weathering process. This includes understanding how biological activity can enhance physical and chemical weathering, leading to soil formation and nutrient cycling.

2. Classification of Sedimentary Rocks

Readers will be able to explain the compositional and generic concepts that apply to the classification of sedimentary rocks, including:

- Clastic Sedimentary Rocks: Understanding the formation of rocks from the accumulation of clasts derived from pre-existing rocks, including sandstone, shale, and conglomerate.

- Chemical Sedimentary Rocks: Recognizing rocks formed from the precipitation of minerals from solution, such as limestone, rock salt, and gypsum.

- Organic Sedimentary Rocks: Identifying rocks formed from the accumulation of organic material, such as coal and certain types of limestone.

3. Physical Characteristics

Readers will describe the physical characteristics that apply to siliciclastic and non-siliciclastic rocks, capturing typical rock types such as:

- Sandstone: Understanding its grain size, sorting, and cementation characteristics.

- Shale: Recognizing its fine-grained texture and fissility.

- Limestone: Identifying its composition, including fossil content and texture variations.

- Coal: Understanding its organic composition and the processes leading to its formation.

4. Physical Properties

Readers will describe the major physical properties of sedimentary rocks as observed in hand specimens, with an emphasis on:

- Texture: Understanding grain size, shape, and distribution.

- Form: Recognizing the overall structure and layering of sedimentary rocks.

- Color Variations: Identifying the significance of color in relation to mineral composition and depositional environment.

5. Sedimentary Structures

Readers will describe major sedimentary structures commonly encountered in the field, focusing on their implications for depositional processes associated with typical rock types, including:

- Cross-Bedding: Understanding its formation and significance in indicating current direction.
- Ripple Marks: Recognizing their formation in shallow water environments and their implications for sediment transport.
- Graded Bedding: Identifying how this structure reflects changes in energy conditions during sediment deposition.

Specific Competences

By the end of this book, readers will be equipped with the abilities to:

1. Distinguish Weathering Features

Readers will be able to identify features produced by physical weathering processes, such as frost wedging, exfoliation, and sheeting, on exposed rock bodies. They will understand the significance of these features in landscape evolution and their implications for soil formation and stability.

2. Chemical Weathering Processes

Readers will demonstrate an understanding of chemical weathering processes, emphasizing the roles of solution, oxidation, and the effects of salts on rock masses. They will learn how these processes contribute to the formation of secondary minerals and influence the overall stability of geological formations.

3. Biological Weathering

Readers will understand the effects of plant roots, burrowing animals, and microbial activities in the biological weathering of rock masses. They will highlight the roles of these organisms in soil formation, nutrient cycling, and the enhancement of weathering processes.

4. Bioturbation

Readers will demonstrate an understanding of bioturbation as undertaken by insects and lichens, and how these processes contribute to sediment mixing and alteration. They will recognize the ecological significance of bioturbation in enhancing soil health and promoting biodiversity.

5. Sedimentary Concepts and Processes

Readers will demonstrate an understanding of sedimentary concepts and processes that transform clastic and non-clastic materials from source rock through diagenetic processes to the formation of bedrock. This includes an understanding of compaction, cementation, and the role of pore fluids in sedimentary rock formation.

6. Distinguish Sedimentary Rocks

Readers will be able to distinguish sedimentary rocks based on a compositional approach, focusing on their environments of deposition, such as fluvial, deltaic, and marine settings. They will learn to interpret sedimentary structures and textures to infer depositional environments.

7. Identify Rock Types

Readers will develop the skills to identify siliciclastic and non-siliciclastic sedimentary rocks in the field based on textural and structural features. They will utilize field observations and hand specimen analysis to classify and describe rock types accurately.

8. Surface Textures of Sediments

Readers will demonstrate an understanding of the surface textures of sediments, with an emphasis on rounding and sphericity typical of dynamic fluvial systems. They will learn how these characteristics reflect transport history and depositional environments, providing insights into past geological processes.

Conclusion

This book serves as a comprehensive resource for understanding sedimentary rocks, their formation processes, and their significance within the geological record. Throughout the chapters, we have explored the intricate relationships between weathering, sedimentation, and diagenesis, highlighting how these processes contribute to the development of various sedimentary rock types. By delving into these interconnections, readers can appreciate the dynamic nature of sedimentary geology and the multifaceted factors that influence the formation and characteristics of these rocks.

By achieving the outlined learning outcomes and competencies, readers will gain a robust foundation in sedimentary geology, enabling them to engage with the subject matter in both academic and practical contexts. This foundational knowledge is essential for students, professionals, and enthusiasts alike, as it equips them with the analytical skills necessary to critically assess sedimentary rocks. Readers will learn to interpret the physical and chemical characteristics of these rocks, understand their depositional environments, and draw conclusions about past geological events and climate conditions. This ability to analyze and interpret sedimentary rocks is crucial for various applications, including resource exploration, environmental assessment, and geological research.

Moreover, this book emphasizes the importance of sedimentary rocks in various fields, including natural resource management, environmental science, and paleoclimatology. Understanding sedimentary processes is essential for addressing contemporary challenges such as resource extraction, land use planning, and environmental conservation. For instance, knowledge of sedimentary rock formations can inform strategies for sustainable resource management, helping to balance economic needs with environmental protection. In the context of fossil fuel extraction, understanding the sedimentary environments that host oil and gas reservoirs can lead to more efficient and environmentally responsible extraction practices.

As readers delve into the dynamic processes that shape our planet's surface, they will foster a deeper appreciation for the complexity and interconnectedness of geological systems. The study of sedimentary rocks not only enriches our understanding of Earth's history but also informs our approach to managing and preserving the natural world. By recognizing the role of sedimentary rocks in the broader context of geological processes, readers can better understand the implications of these formations for current and future environmental challenges. For example, sedimentary rocks play a critical role in carbon sequestration, influencing climate change mitigation efforts.

In addition to the scientific and practical implications, this book aims to inspire a sense of wonder and curiosity about the natural world. The beauty and diversity of sedimentary rocks, from the intricate patterns of sedimentary structures to the vibrant colors of mineral compositions, reflect the complex history of our planet. By fostering an appreciation for these natural wonders, we hope to encourage readers to engage with their surroundings, whether through fieldwork, research, or simply exploring local geological features.

In conclusion, we hope this book inspires curiosity and encourages further exploration of sedimentary geology. By equipping readers with the tools necessary to contribute meaningfully to the field, we aim to foster a new generation of geologists and environmental scientists who are well-prepared to tackle the complexities of our planet's geological processes. We encourage readers to continue their journey of discovery, engaging with the rich tapestry of Earth's geological history and its ongoing evolution. Through this exploration, we can all play a part in enhancing our understanding of the natural world and promoting its sustainable stewardship for future generations.

As we look to the future, the importance of sedimentary geology will only continue to grow. With ongoing challenges such as climate change, resource depletion, and environmental degradation, the insights gained from studying sedimentary rocks will be invaluable. By understanding the processes that shape our planet, we can make informed decisions that benefit both society and the environment. We invite readers to take the knowledge gained from this book and apply it in their pursuits, whether in academia, industry, or conservation efforts, contributing to a more sustainable and informed future for our planet.

CHAPTER ONE

WEATHERING PROCESSES OF SEDIMENTARY ROCKS

1.1. Introduction

Sedimentary rocks are formed from loose fragments of pre-existing natural earth materials that have been weakened and transformed into a coherent mass on the Earth's surface and beneath water bodies. These weakened materials, known as sediments, vary in composition and continue to be generated today. Sediments can originate from weathering processes acting on parent rock masses, after which they are transported to various locations for deposition by agents such as flowing water, wind, glaciers, and gravitational forces. In some cases, sediments may undergo negligible transportation or remain in situ, accumulating in their original locations.

Despite being estimated to constitute only about 8% of the total volume of the Earth's crust, sedimentary rocks extensively cover the crust. This prevalence highlights their significance in geological processes and landscape formation.

Weathering and Erosion

Weathering is the process through which intact rock masses are weakened, leading to the breakdown of these materials into smaller fragments. It is a fundamental process that contributes to the alteration of landscapes across various environments. Erosion, on the other hand, involves the removal and transportation of weakened materials from their original locations. Together, weathering and erosion are key processes in geomorphology, the study of the nature and origin of landforms. The interaction between the atmosphere, hydrosphere, and landmasses results in the creation of new landforms, reflecting the dynamic nature of the Earth's surface.

There are two primary processes of weathering that contribute to the weakening of rock masses: physical and chemical weathering. Physical weathering involves the mechanical breakdown of rocks without altering their chemical composition, often through processes such as freeze-thaw cycles, thermal expansion, and abrasion. Chemical weathering, in contrast, involves the alteration of the mineral composition of rocks through chemical reactions, such as hydrolysis, oxidation, and carbonation.

A complementary process, biological weathering, is facilitated by living organisms, including plants, animals, and microorganisms. This type of weathering can significantly impact rock masses, as seen in adverse agricultural practices that lead to land degradation. The activities of organisms, such as root growth and burrowing, can enhance the weathering process, further contributing to the breakdown of rocks and the formation of sediments.

Fig.3.1: Talus cones, Banff National Park, Alberta

In summary, the interplay of weathering and erosion is crucial in shaping the Earth's landscapes, leading to the formation of sedimentary rocks and influencing the distribution of sediments across various environments. Understanding these processes is essential for comprehending the geological history and evolution of the Earth's surface.

1.2 Physical Weathering

Physical weathering refers to the disintegration or degradation of rocks without any chemical alteration. This process involves various mechanisms that break down rocks into smaller fragments while maintaining their original mineral composition. Key processes of physical weathering include frost wedging, exfoliation, sheeting, and volume changes in affected minerals.

Freezing and Thawing

One of the most significant processes of physical weathering is frost action, which involves volumetric expansion occurring in closed voids within rocks that are nearly saturated with water. This phenomenon is most effective when temperatures oscillate between -3°C and -8°C. The process, commonly known as frost wedging, occurs when water infiltrates cracks in the rock and subsequently freezes. As water freezes, it expands by approximately 9%, exerting considerable strain on the surrounding rock. The stress exerted by the frozen water can reach about 550 psi, a value comparable to the tensile strength of many rock types.

Repeated cycles of freezing and thawing subject the affected rocks to significant pressure, ultimately leading to their breakdown. The landforms created as a result of this oscillatory effect include talus cones and scree slopes, which are accumulations of rock debris at the base of cliffs. Figure 3.1 illustrates typical talus cones formed through this process.

Differential Expansion and Contraction

Another important mechanism of physical weathering is thermal expansion and contraction, which arises from the repeated daily heating and cooling of minerals within rock masses. When minerals are heated, they expand; conversely, when they cool, they contract. Different minerals expand and contract at varying rates, leading to the development of stresses along mineral boundaries. This differential expansion can weaken the rock and release subsurface confining stress.

As residual stresses accumulate, they can cause vertical fracturing of the rock mass into blocks through a process known as exfoliation (see Fig. 3.2 (A)). If fractures develop in a horizontal direction during this process, it is referred to as sheeting. Notable examples of exfoliation can be observed in the North Dome and Half Dome igneous intrusions located in Yosemite National Park (see Fig. 3.2 (B)). These iconic landforms exhibit the characteristic features of exfoliation, showcasing the effects of thermal expansion and contraction on rock masses.

Fig.3.2: Exfoliation of Granite in Carolina (A) and Sheeting of Granite in Yosemite Park in (B)

Conclusion

Physical weathering plays a crucial role in shaping the Earth's landscape by breaking down rocks into smaller fragments through mechanical processes. Understanding these processes, such as frost wedging and thermal expansion, is essential for comprehending the dynamics of rock weathering and the formation of various landforms. The interplay of these physical weathering mechanisms contributes significantly to the ongoing evolution of the Earth's surface.

Other Agents of Physical Weathering

In addition to frost wedging and thermal expansion, other agents of physical weathering contribute to the mechanical weakening of rock masses. Two significant agents are wind and wind-driven rain, both of which can have profound effects on exposed rock surfaces.

Wind Action

Wind loading exerts pressure on rock exposures, particularly in arid and semi-arid environments where vegetation is sparse (Fig. 3.3). The force of the wind can act on openings and fractures in the rock, leading to the weakening of structural bonds among the mineral constituents. Over time, this mechanical action can cause the disintegration of rock surfaces, particularly those that are already compromised by other weathering processes.

Fig.3.3: Effects of Wind Load on a Building

Wind can also transport fine particles, such as sand and silt, which can act as abrasives against rock surfaces. This process, known as abrasion, further contributes to the physical weathering of rocks. The cumulative effect of wind erosion can lead to the formation of unique landforms, such as ventifacts—rocks that have been shaped and polished by wind-driven particles.

Wind-Driven Rain

Wind-driven rain is another critical agent of physical weathering. When rain is propelled by strong winds, it can penetrate rock exposures more effectively than still rain. This penetration exerts compressive forces on the rock, which can lead to the weakening of the rock structure. The impact of wind-driven rain can be particularly damaging to man-made structures, as seen in historical buildings that have endured prolonged exposure to these elements.

For example, the 16th-century monastery in Mexico (see Fig. 3.3) illustrates the effects of weathering primarily caused by wind and wind-driven rain. The volcanic tuff rock material used in the construction of the monastery has suffered significant degradation over the centuries due to these weathering processes. The visible ravages of weathering on the structure highlight the importance of understanding the impact of environmental factors on both natural and human-made rock formations.

1.3 Chemical Weathering

Chemical weathering refers to the suite of chemical processes that lead to the dissolution and decay of earth materials. Unlike physical weathering, which breaks rocks into smaller pieces without altering their chemical composition, chemical weathering involves the transformation of minerals through chemical reactions. For chemical weathering to occur, rocks must be exposed to water, oxygen, hydrogen, mild acids, and elevated temperatures, which facilitate the necessary reactions.

The process of chemical weathering can be significantly enhanced by mechanical weathering, which increases the surface area of rock blocks. As rocks are broken down into smaller fragments, more surface area is exposed to chemical agents, accelerating the weathering process.

Typical Processes of Chemical Weathering

1. **Solution**: In this process, rocks react with water, gases, and solutions, which may be acidic. The chemical reactions result in the addition or removal of elements from the affected minerals. Several common minerals dissolve in water through the solution process, also referred to as dissolution. For example, halite (rock salt) and calcite (found in limestone and marble) are highly soluble in water.

Limestone and marble, which contain calcite, are particularly susceptible to dissolution in acidic water. Rainwater, which absorbs carbon dioxide from the atmosphere, can become slightly acidic due to the formation of carbonic acid. The reaction can be represented as follows:

CO_2 + H_2O = H_2CO_3

Carbon dioxide water Carbolic acid

Carbonic acid dissociates to produce hydrogen and bicarbonate ions as follows:

H_2CO_3 <=> H^+ + HCO_3^-

Carbolic acid hydrogen ion bicarbonate ion

Typical reaction for chemical weathering of silicate minerals by carbonic acid;

$2KAlSiO_3O_8$ + H2O + 2H2CO3 = $AlSi_2O_5(OH)_2 + 2H^+ + 4SiO_2 + 2$ HCO_3^-

Feldspar water carbonic acid clay minerals dissolved ions

Typical Minerals Involved in Weathering

Weathering processes involve a variety of minerals, each contributing differently to the breakdown and alteration of rocks. Among the most significant minerals participating in weathering are ferromagnesian minerals, feldspars, and quartz.

1. **Ferromagnesian Minerals**: These minerals, which include biotite, hornblende, and olivine, are rich in iron and magnesium. During chemical weathering, ferromagnesian minerals tend to break down and alter to form clay minerals. The weathering of these minerals often results in the generation of highly insoluble iron oxides, particularly varieties of limonite, which manifest as rust-colored deposits. This transformation not only contributes to soil formation but also affects the color and composition of the surrounding rock material.

2. **Feldspars**: Feldspar minerals, such as orthoclase and plagioclase, are among the most abundant minerals in the Earth's crust. They weather through hydrolysis, releasing cations such as sodium, calcium, and potassium into solution. The breakdown of feldspars leads to the formation of clay minerals, including kaolinite (an aluminum silicate), illite, and montmorillonite. These clays play a crucial role in soil development and can influence the physical and chemical properties of the soil.

3. **Quartz**: Unlike ferromagnesian minerals and feldspars, quartz is highly resistant to chemical weathering. It remains chemically unaffected and retains its original composition throughout the weathering process. This durability makes quartz a common component of sedimentary rocks, such as sandstone, and contributes to the stability of soils formed from weathered materials.

Karsts

Karst landscapes are distinctive geological formations that develop primarily in limestone regions where the concentration of carbonate minerals is at least 50%. The formation of karsts is influenced by several factors:

(a) **Permeability and Porosity**: The ability of limestone to transmit water (permeability) and the presence of void spaces within the rock (porosity) are critical for the development of karst features. High permeability allows for the infiltration of water, which can dissolve the carbonate minerals, leading to the formation of caves and other karst features.

(b) **Bedding Thickness**: The thickness of the limestone beds can affect the development of karst landscapes. Thicker beds may provide more substantial material for dissolution, while thinner beds may lead to more localized karst features.

(c) **Available Relief**: The topography of the area plays a significant role in karst development. Regions with significant relief can enhance water flow and erosion, promoting the dissolution of limestone and the formation of karst features such as sinkholes and underground rivers.

(d) **Secondary Porosity**: The presence of fractures, joints, and faults in the limestone can create secondary porosity, allowing water to penetrate deeper into the rock. This increased water movement can accelerate the dissolution process and lead to the development of extensive cave systems.

One of the most extensive and well-developed karst landscapes is referred to as "tower karst," which is prominently found in the Guangxi province of China (see Fig. 3.4). In this region, towering limestone landforms are characterized by steep cliffs and pinnacles, often riddled with relict caves at high elevations and active caves at lower levels. The unique topography and geological features of tower karst landscapes provide valuable insights into the processes of karstification and the dynamic interactions between water and limestone.

Conclusion

The interplay between various minerals and the processes of weathering significantly shapes the Earth's surface and contributes to the formation of distinctive geological features such as karsts. Understanding the characteristics and behaviors of minerals involved in weathering, as well as the factors influencing karst development, is essential for comprehending the complexities of geological processes and landscape evolution.

Fig.3.4: Karst Topography, Guilin Region, China

Cave Deposits and Speleothems

Caves, formed primarily through the dissolution of soluble rocks such as limestone, often become partially or completely infilled with various deposits over time. Among the most notable of these deposits are stalactites and stalagmites, which are collectively referred to as speleothems. These fascinating geological formations are generated by the action of percolating water that carries dissolved minerals into the cave environment.

Formation of Speleothems

The process of speleothem formation begins when rainwater, which has absorbed carbon dioxide from the atmosphere and soil, seeps into the ground. As this water moves through the soil and rock layers, it becomes slightly acidic due to the presence of carbonic acid formed from the dissolved carbon dioxide. This acidic water is capable of dissolving calcium carbonate ($CaCO_3$) from the limestone, leading to the formation of calcium bicarbonate in solution:

Carbonic acid dissociates to produce hydrogen and bicarbonate ions as follows:

H_2CO_3	<=>	H^+	+	HCO_3^-
Carbolic acid		hydrogen ion		bicarbonate ion

When this mineral-rich water reaches the cave atmosphere, which typically has lower levels of carbon dioxide, it undergoes a process known as degassing. As the carbon dioxide escapes from the solution, the water becomes supersaturated with calcium carbonate. This supersaturation triggers the precipitation of calcium carbonate, leading to the formation of speleothems.

Types of Speleothems

1. **Stalactites**: These formations hang from the ceilings of caves and are created by the dripping of mineral-laden water from above. As the water drips, it leaves behind a small deposit of calcium carbonate, gradually building up over time. Stalactites typically taper to a point and can vary in size from a few centimeters to several meters in length.

2. **Stalagmites**: In contrast to stalactites, stalagmites grow upward from the cave floor. They are formed by the accumulation of calcium carbonate from dripping water that falls to the ground. Stalagmites can also vary in size and shape, often appearing as rounded or conical structures.

3. **Other Speleothem Features**: In addition to stalactites and stalagmites, a variety of other speleothem formations can develop within caves, including columns (which form when a stalactite and stalagmite grow together), flowstones (thin sheets of mineral deposits that form on cave floors or walls), and draperies (thin, curtain-like formations that hang from cave ceilings).

Environmental Conditions and Speleothem Deposition

The environmental conditions within caves play a significant role in the formation and deposition of speleothems. Near the entrances of many caves, evaporation of drip waters is often heightened due to increased airflow and exposure to external conditions. This evaporation enhances the deposition of calcium carbonate, leading to the accumulation of calcitic features at the entrances of caves. As a result, many caves exhibit a characteristic "choked" appearance at their entrances, where stalagmites and other calcitic formations dominate the landscape (see Fig. 3.5).

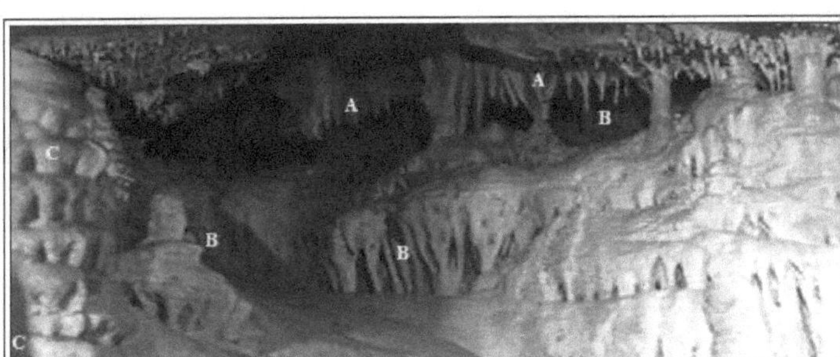

Fig. 3.5: Typical Caves with Stalactites(A), Stalagmites(B) and Flow wall (C)

Speleothems: Insights into Geological Processes

Speleothems, which include stalactites, stalagmites, and other mineral formations found within caves, are remarkable geological features that serve as indicators of the complex processes involved in cave formation and the interactions between water and rock. These formations are not merely aesthetic; they provide critical insights into the environmental conditions and geological history of karst landscapes.

Formation of Speleothems

The formation of speleothems is a multi-step process that begins with the dissolution of soluble rocks, primarily limestone, by slightly acidic water. As rainwater percolates through soil and rock layers, it absorbs carbon dioxide, forming carbonic acid. This acidic solution is capable of dissolving calcium carbonate from limestone, leading to the creation of calcium bicarbonate in solution. When this mineral-rich water enters a cave, it encounters a lower concentration of carbon dioxide, causing it to degas and become supersaturated with calcium carbonate. This supersaturation results in the precipitation of calcium carbonate, leading to the gradual formation of speleothems.

The growth of stalactites and stalagmites occurs over thousands to millions of years, with the rate of deposition influenced by various factors, including the rate of water flow, the concentration of dissolved minerals, and the cave's microclimate. For instance, a cave with a high rate of water dripping will typically produce larger and more rapidly growing stalactites and stalagmites compared to a cave with minimal water flow.

Indicators of Environmental Change

Speleothems are not only products of geological processes but also serve as valuable indicators of past environmental conditions. The isotopic composition of the calcium carbonate in speleothems can provide insights into historical climate conditions, such as temperature and precipitation patterns. For example, variations in oxygen and carbon isotopes within the speleothem layers can reflect changes in the hydrological cycle and climate over time. This makes speleothems important tools for paleoclimatology, allowing scientists to reconstruct past climates and understand how they have influenced the development of karst landscapes.

Karst Environments and Their Dynamics

The presence of speleothems highlights the dynamic nature of karst environments, which are characterized by unique geological features formed through the dissolution of soluble rocks. These environments are shaped by a delicate balance of chemical processes, hydrology, and geological factors. The interplay between water movement, mineral dissolution, and deposition creates a complex system that continually evolves over time.

Karst landscapes often feature a variety of landforms, including sinkholes, underground rivers, and extensive cave systems. The formation of speleothems within these caves is a reflection of the ongoing geological processes that shape the landscape. Understanding these processes is crucial for managing and preserving karst environments, which are often sensitive to human activities such as mining, agriculture, and urban development.

Broader Geological and Hydrological Implications

The study of speleothems extends beyond cave systems and contributes to broader geological and hydrological studies. By examining the formation and characteristics of speleothems, researchers can gain insights into groundwater flow patterns, mineral deposition processes, and the overall health of karst aquifers. This knowledge is essential for water resource management, particularly in regions where karst aquifers serve as important sources of drinking water.

Moreover, speleothems can provide information about the geological history of an area, including tectonic activity, erosion rates, and sedimentation processes. This information is valuable for understanding the evolution of landscapes and the factors that have shaped them over geological time scales.

Conclusion

In summary, speleothems, including stalactites and stalagmites, are not only fascinating geological formations but also critical indicators of the processes that govern cave formation and the interactions between water and rock. Their study enhances our understanding of karst environments and contributes to broader geological and hydrological research. By unraveling the complexities of speleothem formation and their implications for past and present environmental conditions, we can better appreciate the intricate relationships that define our planet's geological landscape.

Salt Crystallization and Weathering: Salt crystallization is a significant geological process that occurs when saline solutions evaporate, leading to the formation of salt crystals. This process is particularly relevant in arid and semi-arid environments, where the conditions are conducive to the accumulation and subsequent weathering of salt minerals. The mechanisms involved in salt weathering can have profound effects on rock masses, contributing to their mechanical weakening and alteration (Fig.3.6).

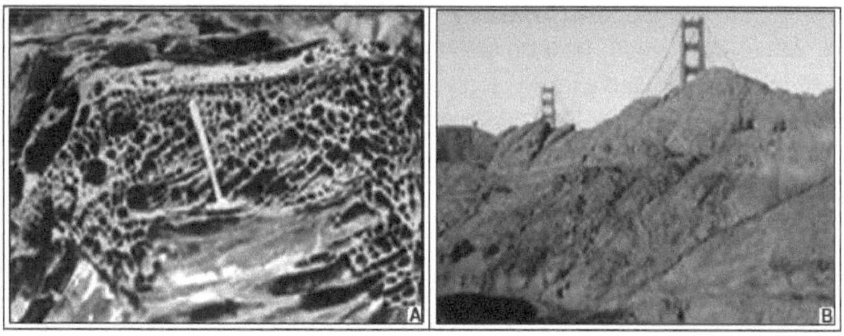

Fig.3.6: Honeycomb features(A) and Honeycomb weathering features (B) in Greywacke Sandstone of Golden Date Recreation Area, San Fransisco California

Mechanism of Salt Weathering

When saline solutions evaporate, the dissolved salts precipitate out of the solution, forming solid crystals. These crystals can include common minerals such as halite (sodium chloride), gypsum (calcium sulfate), and various other salts. As these crystallized minerals absorb moisture from the environment, they can dissolve back into solution, leading to a cycle of crystallization and dissolution. This process generates expansion as the crystals grow, exerting significant force on the surrounding rock mass. The mechanical stress induced by this expansion can lead to the fracturing and disintegration of the rock, contributing to the overall weathering process.

Environmental Conditions Favoring Salt Weathering

Salt weathering is particularly prevalent in two types of environments:

1. **Arid and Semi-Arid Regions**: In these areas, capillary rise can bring saline groundwater to the surface, where it evaporates, leaving behind salt deposits. The high evaporation rates in these regions promote the concentration of salts, which can then undergo repeated cycles of crystallization and dissolution. Additionally, the large diurnal temperature

fluctuations common in hot arid regions can lead to significant changes in relative humidity, further enhancing the wetting and drying cycles that facilitate salt weathering.

2. **Cold Regions**: In colder climates, slight increases in temperature can encourage the precipitation of salts from salt-saturated water bodies. During warmer periods, the solubility of salts can decrease, leading to crystallization. This process can occur in environments such as saline lakes or coastal areas where evaporation rates are high during warmer months.

Ecological Implications of Salt Weathering

The empty spaces created by salt weathering are numerous and often exposed to the atmosphere. These voids can act as traps for airborne seeds, spores, and other microorganisms. As these biological entities settle into the crevices and cavities formed by salt weathering, they can begin to grow and flourish. The presence of vegetation and microbial life contributes to further chemical and biological weathering of the surrounding rock.

The growth of plants in these environments can enhance weathering through several mechanisms:

- **Root Expansion**: Plant roots can penetrate cracks and voids in the rock, exerting physical pressure that contributes to rock disintegration. As roots grow, they can also secrete organic acids that chemically weather the rock material.

- **Microbial Activity**: Microorganisms, including bacteria and fungi, can produce organic acids and other metabolites that facilitate the breakdown of minerals. This biological weathering process can lead to the release of nutrients, further promoting plant growth and contributing to the development of soil.

- **Soil Formation**: Over time, the accumulation of organic matter from decaying plants and microorganisms can lead to the formation of soil in these previously barren environments. This soil can support a wider range of vegetation, further enhancing the weathering processes.

Conclusion

Salt crystallization and weathering are important geological processes that significantly impact rock masses, particularly in arid and semi-arid environments.

The cycles of crystallization and dissolution, driven by environmental conditions, contribute to the mechanical weakening of rocks and the formation of voids that can support biological life. Understanding the mechanisms and implications of salt weathering not only enhances our knowledge of geological processes but also highlights the intricate relationships between geology, ecology, and the environment. As these processes unfold, they contribute to the ongoing evolution of landscapes and ecosystems, demonstrating the dynamic nature of the Earth's surface.

Over the past decade, there has been a growing scientific interest in the phenomena of building material decay caused by salt weathering. This decay is increasingly recognized as one of the primary agents responsible for the deterioration of historical architecture, structures at archaeological sites, and various archaeological objects. The impact of salt weathering on cultural heritage is profound, as it can lead to significant structural damage, loss of aesthetic value, and the degradation of historical integrity.

The Importance of Understanding Salt Weathering

Concerted research efforts are currently underway to better understand the mechanisms of salt weathering and to develop effective strategies for mitigating the damage it causes. This research is crucial for the preservation of cultural heritage structures, as it aims to identify the specific salts involved, the conditions that promote their crystallization, and the best conservation measures to protect these invaluable assets. By addressing the challenges posed by salt weathering, conservationists can implement more effective restoration techniques and prolong the lifespan of historical buildings and artifacts.

Rates of Chemical Weathering

The rates of chemical weathering vary significantly among different minerals, and past studies have provided valuable insights into these processes. Research on the disintegration of various silicate minerals has shown that the rate of chemical weathering is inversely related to Bowen's Reaction Series (Fig.3.7). This series categorizes minerals based on their temperature of formation, with those formed at higher temperatures being less stable at the Earth's surface.

Bowen's Reaction Series and Mineral Stability

According to Bowen's Reaction Series, minerals that crystallize at higher temperatures, such as olivine and pyroxene, are more susceptible to weathering when exposed to surface conditions. For instance, olivine minerals weather most rapidly compared to other silicate minerals. This rapid weathering is attributed to their relatively unstable structure when subjected to the lower temperatures and varying chemical conditions found at the Earth's surface.

In contrast, quartz, which is found at the lower end of Bowen's Reaction Series, exhibits remarkable stability. As free silica, quartz retains its chemical integrity and is highly resistant to weathering processes. The strong covalent bonding within its tetrahedral framework structure contributes to its durability, allowing quartz crystals to withstand attacks from acidic solutions and other weathering agents. This resistance to disintegration makes quartz a common component of sedimentary rocks and a significant contributor to the stability of soils formed from weathered materials.

Conclusion

The increasing recognition of salt weathering as a significant factor in the decay of building materials underscores the need for ongoing research and conservation efforts. Understanding the rates of chemical weathering and the stability of various minerals, as illustrated by Bowen's Reaction Series, is essential for developing effective strategies to protect cultural heritage structures. By addressing the challenges posed by salt weathering and other weathering processes, researchers and conservationists can work together to preserve the integrity and longevity of historical architecture and archaeological sites for future generations. The interplay between geological processes and human-made structures highlights the importance of interdisciplinary approaches in the fields of geology, conservation science, and cultural heritage management.

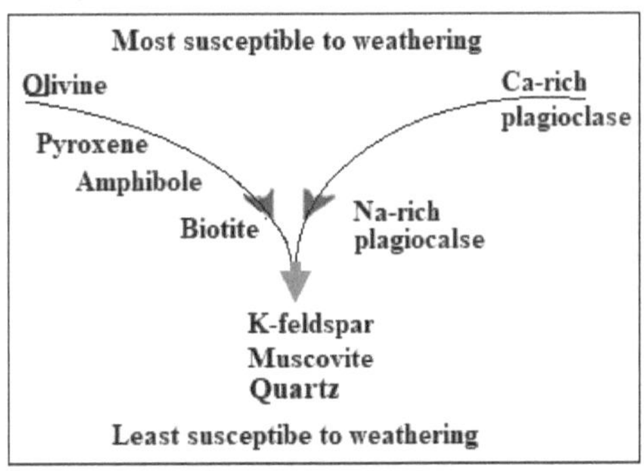

Fig. 3.7: Bowen's Reaction Series for Weathering Susceptibility

Hydrolysis: Hydrolysis involves dissociation of water into hydrogen cation (H^+) and hydroxyl anion (OH^-). H^+ generated displaces other cations in mineral structures. Cations particularly, K^+, Na^+, Ca^{2+}, Mg^{2+} may combine with hydroxyl anion or be carried away in solution. The process of Hydrolysis is promoted by;

(a) Decreasing pH (increasing H^+),

(b) Decomposition of organic matter (releases H^+),

(c) Increase in water temperature.

A typical mineral reaction i.e. hydrolysis that involves albite and water is in the form;

$$4NaAlSi_3O_8 + 6H2O = Al_4Si_4O_{10}(OH)_8 + 8SiO_2 + 4Na^+ + 4OH^-$$

Albite water kaolinite silica sodium ion hydroxyl ion

Thus, the albite weathers to form kaolinite and produces some residual silica. Similarly, alkali feldspar produces clay minerals (Kaolinite) as follows:

$$KAlSi3O8 + H2O \Rightarrow Al2Si2O5(OH)4 + SiO2 + K^+$$

Feldspar water
kaolinite silica
potassium ions

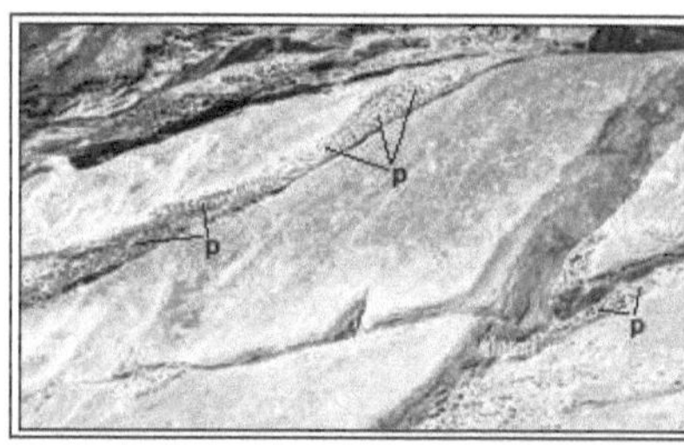

Fig.3.8: Pockmarks(P) on exposed Granite Cape Sebastian State Park Oregon USA

Weathering Features of Granitic Intrusive Bodies: Granitic intrusive bodies, which are formed from the slow crystallization of magma beneath the Earth's surface, commonly exhibit a variety of weathering features (Fig.3.8). These features are indicative of the complex interactions between geological materials and environmental conditions. Among the most notable weathering features found in granitic intrusions are exfoliations, sheetings, and pockmarks.

Exfoliation and Sheeting

Exfoliation is a physical weathering process characterized by the peeling or flaking of rock layers from the surface of a granitic body. This phenomenon occurs due to the release of pressure as overlying materials are eroded away, allowing the rock to expand and fracture along parallel planes. The result is a series of thin, curved layers that can resemble the layers of an onion.

Sheeting, closely related to exfoliation, involves the separation of rock layers along natural fractures or joints. This process is often facilitated by thermal expansion and contraction, where temperature fluctuations cause the rock to expand during the day and contract at night. Over time, these repeated cycles can lead to the development of large, sheet-like slabs of rock that detach from the main body.

Pockmarks

Pockmarks are another distinctive weathering feature commonly observed in granitic intrusive bodies. These rounded depressions are prevalent in exposed rock surfaces and can be seen in various geological settings (see Fig. 3.8). Pockmarks are formed through a combination of chemical and physical weathering processes, where certain minerals within the granite weather more rapidly than others.

The formation of pockmarks begins when water collects in small depressions or pits on the rock surface. As the water interacts with the minerals, it can lead to enhanced weathering, particularly in areas where softer or more reactive minerals, such as feldspar, are present. Over time, the continuous action of water and weathering processes causes these depressions to enlarge. As they grow, pockmarks can merge with adjacent depressions, creating a distinctive "pock-marked" surface that is characteristic of weathered granitic rock.

Ecological Implications

The pockmarked surfaces of granitic intrusions provide unique ecological niches that can support the growth of various organisms. The depressions created by pockmarks can collect moisture, creating microhabitats that are conducive to the establishment of lichen, mosses, and even small plants. These organisms play a vital role in the ecosystem by contributing to soil formation, enhancing nutrient cycling, and providing habitat for other species.

The presence of vegetation in these pockmarked areas can further influence weathering processes. As plants grow, their roots can penetrate the depressions, exerting physical pressure on the surrounding rock and promoting additional weathering. Moreover, the organic acids produced by plant roots and microbial activity can enhance chemical weathering, leading to further alteration of the rock material.

Conclusion

Granitic intrusive bodies exhibit a range of weathering features, including exfoliations, sheetings, and pockmarks, each reflecting the dynamic processes of weathering and erosion. Pockmarks, in particular, are significant as they illustrate the interplay between chemical and physical weathering, resulting in distinctive depressions that contribute to the ecological diversity of the area. Understanding these weathering features not only enhances our knowledge of geological processes but also highlights the importance of these formations in supporting various forms of life and contributing to the overall health of the ecosystem. The study of weathering in granitic intrusions provides valuable insights into the long-term evolution of landscapes and the interactions between geological and biological systems.

Oxidation in Geological Processes

Oxidation is a critical chemical weathering process that typically occurs in a water-free environment, preferably above the water table. This process involves elements within a mineral structure losing electrons, which results in an increase in their ionic charge. The loss of electrons is primarily due to the reaction of metal compounds with oxygen ions (O^{2-}) or hydroxide anions (OH^-). The oxidation process can significantly alter the mineral composition of rocks and contribute to the overall weathering of geological materials.

Mechanisms of Oxidation

1. **Reaction with Oxygen Ions (O^{2-})**: When minerals containing metallic elements are exposed to oxygen, oxidation can occur. For example, iron-bearing minerals such as olivine, pyroxene, and biotite can react with oxygen to form iron oxides. This reaction can be represented as follows:

(a) $4Fe^{+2} + 3O_2 = 2Fe_2O_3$

Ferrous iron (Fe^{+2})	oxygen	Ferric iron (Fe^{+3})
e.g. in iron		e.g. hematite

(b) Olivine weathers through the combination of hydrolysis and oxidation to form Hematite, Thus;

$2Fe_2SiO_4 + 4H_2O + O_2 = 2Fe_2O_3 + 2H_4SiO_4$

Olivine	water	oxygen	hematite	silicic acid

2. **Reaction with Hydroxide Anions (OH^-)**: Oxidation can also occur through reactions involving hydroxide ions. For instance, when iron-bearing minerals react with hydroxide ions in the presence of water, they can form iron hydroxides, which may further oxidize to form stable iron oxides. This iron hydroxide can then oxidize to form iron oxides such as goethite ($FeO(OH)$) or hematite (Fe_2O_3).

Oxidation is a significant chemical weathering process that occurs when reducing agents react with oxygen to form cations such as Fe^{2+}, Al^{3+}, Mg^{2+}, Na^+, Ca^{2+}, and K^+. Among the various elements that undergo oxidation, iron is the most common oxidized material found at the Earth's surface. Most elements in this environment exist in an oxidized state, which plays a crucial role in the weathering of minerals and the alteration of rock formations.

Iron-Bearing Silicate Minerals

Typical iron-bearing silicate minerals that combine with oxygen to produce iron oxides include olivine, pyroxene, amphibole, and biotite. When these minerals oxidize, they can form secondary minerals such as limonite, hematite, and goethite. The iron oxides produced through these reactions are commonly characterized by their red, orange, or brown colors, which can significantly influence the appearance of weathered rocks.

For example, mafic rocks such as basalt, which may contain olivine, pyroxene, or amphibole, often weather to an orange hue due to the oxidation of iron-bearing minerals. This color change is a visible indicator of the chemical processes occurring within the rock.

Reduction and Its Effects

Reduction, the opposite of oxidation, occurs in environments below the water table, where oxygen levels are lower. In these reducing conditions, the reduced forms of elements are generally more mobile and soluble. This mobility can lead to the leaching of essential nutrients and minerals from the rock, further altering its composition and structure.

Many metallic ores and minerals also undergo oxidation and dissolution, contributing to the coloration of deposits. A typical example is chalcopyrite ($CuFeS_2$), which oxidizes to form copper hydroxide and iron oxides. This transformation not only affects the mineral's appearance but also its economic viability as a source of copper.

Chemical Weathering Processes Resulting from Oxidation

Chemical weathering processes associated with oxidation lead to several distinct features and phenomena:

1. **Spalling**: This process refers to the formation of chips, fragments, or flakes from a piece of stone as a result of chemical weathering. Spalling can occur when the outer layers of a rock expand and contract due to temperature changes or moisture fluctuations, causing them to break away.

2. **Weathering Rinds**: Weathering rinds are the outer layers of pebbles, boulders, or other rock fragments that form as a result of chemical weathering. These rinds often exhibit different colors and textures compared to the unweathered interior of the rock, reflecting the chemical changes that have occurred.

3. **Spheroidal Weathering**: In spheroidal weathering, exfoliation caused by chemical weathering results in the disaggregation of a rock, progressively forming spherical features. This process is particularly evident in granitic rocks, where the weathering occurs along joints and fractures. Over time, this leads to the generation of rounded boulders, which are characteristic of weathered granite landscapes.

4. **Clay Mineral Production**: The disintegration of alkali feldspars during chemical weathering often results in the formation of secondary minerals, particularly kaolinite. The generation of clay minerals contributes to the overall alteration of the rock and can enhance its porosity, allowing for increased fluid permeation. This, in turn, accelerates the chemical weathering processes affecting the rock mass.

Conclusion

Oxidation plays a critical role in the chemical weathering of rocks, particularly those containing iron-bearing minerals. The processes of oxidation and reduction significantly influence the stability and appearance of minerals, leading to various weathering features such as spalling, weathering rinds, and spheroidal weathering. Understanding these processes is essential for comprehending the long-term evolution of landscapes and the interactions between geological materials and environmental conditions. The study of oxidation and its effects on rock weathering not only enhances our knowledge of geological processes but also informs conservation efforts and resource management in areas affected by weathering phenomena.

Weathering Characteristics of Magmatic Granites in Kenya

In the western and Nyanza regions of Kenya, magmatic granites are prevalent, particularly within the geological framework of the Tanzania Craton. These granitic formations exhibit weathering characteristics that are comparable to those observed in similar geological environments across the globe, including regions in Australia and the United States (see Fig. 3.9).

Fig.3.9: Spheroidal weathering of Granite in Alabama(A), Australia(B) and in Bungoma, Kenya(C)

Geological Context

The Tanzania Craton is one of the oldest geological formations in East Africa, characterized by its stable continental crust and extensive exposure of ancient rocks, including granites. The magmatic granites found in this region are primarily formed from the slow cooling and solidification of magma beneath the Earth's surface. Over time, these granitic bodies have been subjected to various weathering processes that have shaped their current form and composition.

Weathering Processes

The weathering characteristics of magmatic granites in Kenya can be attributed to several key processes:

1. **Physical Weathering**: Physical weathering processes, such as exfoliation and thermal expansion, are common in granitic environments. The release of pressure as overlying materials are eroded can lead to the peeling of rock layers, while temperature fluctuations can cause expansion and contraction, resulting in the fracturing of the rock.

2. **Chemical Weathering**: Chemical weathering plays a significant role in the alteration of granitic rocks. The presence of water, oxygen, and organic acids can lead to the breakdown of feldspar minerals into clay minerals, such as

kaolinite. Oxidation of iron-bearing minerals can also contribute to the development of secondary minerals and the characteristic coloration of weathered granites.

3. **Biological Weathering**: The influence of biological activity, including the growth of plants and microorganisms, can enhance weathering processes. Roots can penetrate cracks in the granite, exerting physical pressure and contributing to the disintegration of the rock. Additionally, organic acids produced by decaying plant material can further promote chemical weathering.

Comparative Analysis

The weathering characteristics of magmatic granites in Kenya are similar to those observed in other regions with comparable geological settings, such as Australia and the United States. In these areas, granitic rocks also exhibit features such as spheroidal weathering, weathering rinds, and the formation of pockmarks. The similarities in weathering processes can be attributed to the shared environmental conditions, including climate, vegetation, and geological history.

For instance, in Australia, granitic landscapes are often characterized by rounded boulders and extensive weathering features resulting from both physical and chemical processes. Similarly, in the United States, granitic formations such as those found in Yosemite National Park display distinct weathering patterns that reflect the interplay of environmental factors.

Conclusion

The magmatic granites of western and Nyanza Kenya, situated within the Tanzania Craton, exhibit weathering characteristics that are consistent with those found in similar geological environments worldwide. Understanding these weathering processes is essential for comprehending the geological history of the region, as well as for informing conservation and land management practices. The comparative analysis of weathering in different regions highlights the universal nature of geological processes while also emphasizing the unique environmental factors that shape each landscape. Continued research into the weathering characteristics of these granitic formations will enhance our understanding of their evolution and the broader geological context in which they exist.

1.4 Biological Weathering

Biological weathering is a significant process that contributes to the breakdown and alteration of rocks and minerals, primarily through the activities of living organisms (Fig.6.10). The types of weathering processes that occur at any given location are largely influenced by the climate of the region. For example, physical weathering processes tend to dominate in cold and dry climates, while chemical weathering is more prevalent in warm and humid environments. In tropical regions, where high temperatures and humidity accelerate chemical reactions, biological weathering plays a crucial role in the disintegration of rocks.

Fig.3.10: Roots Wedging in Rock Outcrops of Maryland(A) and Texas Olstead Island(B)

The Role of Living Organisms

Biological weathering involves various living organisms, including vegetation, animals, and microorganisms, which are particularly active in warm and humid conditions. In these climates, rocks can disintegrate into sediments that provide habitats for microorganisms. These microorganisms contribute to biological weathering both on land and in marine environments, where they can influence sediment composition and rock stability.

Biological weathering often occurs as a combination of chemical and mechanical weathering processes. The interactions between living organisms and geological materials can lead to significant changes in the physical and chemical properties of rocks.

Plant Roots and Their Impact

One of the most notable agents of biological weathering is plant roots. As plants grow, their roots penetrate the soil and can extend into cracks and joints in the underlying rock. This process has several important implications for weathering:

1. **Mechanical Weathering**: As plant roots grow, they exert pressure on the surrounding rock. This pressure can lead to the physical disintegration of

the rock mass, particularly in areas where the rock is already fractured. Over time, the mechanical force exerted by the roots can cause the rock to break apart, creating more space for the roots to expand and for other weathering processes to occur.

2. **Chemical Weathering**: In addition to mechanical effects, plant roots can also contribute to chemical weathering. Many plants produce organic acids as part of their metabolic processes. These organic acids can help dissolve minerals in the rock and soil, facilitating the breakdown of the rock material. For example, citric acid and oxalic acid produced by plant roots can react with minerals such as calcium carbonate, leading to their dissolution.

3. **Moisture Retention**: The growth of plant roots through joints and cracks in the rock allows for the accumulation of moisture in these spaces. This moisture can enhance chemical weathering processes, as it provides the necessary medium for chemical reactions to occur. Additionally, the presence of moisture can support the growth of microorganisms, further contributing to biological weathering.

4. **Nutrient Cycling**: As rocks break down into sediments, they release essential nutrients that can be utilized by plants and microorganisms. This nutrient cycling is vital for the health of ecosystems, as it supports plant growth and contributes to soil formation.

Microbial Activity

Microorganisms, including bacteria and fungi, also play a significant role in biological weathering. These organisms can colonize rock surfaces and contribute to weathering through various mechanisms:

- **Biofilm Formation**: Microorganisms can form biofilms on rock surfaces, which can trap moisture and organic matter. This biofilm can create a microenvironment that enhances chemical weathering processes.

- **Organic Acid Production**: Many microorganisms produce organic acids as metabolic byproducts. These acids can dissolve minerals and contribute to the breakdown of rock material, similar to the effects of plant root exudates.

- **Bioturbation**: The activities of burrowing animals can also contribute to biological weathering. As animals dig and move through the soil, they can disturb rock and sediment, exposing new surfaces to weathering processes.

Conclusion

Biological weathering is a vital process that significantly influences the breakdown and alteration of rocks, particularly in warm and humid climates. The activities of plants, animals, and microorganisms contribute to both mechanical and chemical weathering, leading to the disintegration of rocks into sediments that support diverse ecosystems. Understanding the mechanisms of biological weathering is essential for comprehending the interactions between living organisms and geological materials, as well as the broader implications for soil formation, nutrient cycling, and ecosystem health. The interplay between biological and geological processes highlights the dynamic nature of the Earth's surface and the importance of preserving these interactions for environmental sustainability.

Soil Formation and Biochemical Activities of Animals

Soil formation is a complex process that is significantly enhanced by the biochemical activities of various organisms, ranging from microscopic single-celled organisms to larger mammals that create temporary or permanent homes in the soil. These animals play a crucial role in the development and health of soil ecosystems, particularly through their burrowing activities, which occur before the induration processes transform sediments into bedrock.

The Role of Earthworms

One of the most well-known soil organisms is the earthworm, specifically *Lumbricus terrestris*. Earthworms are vital contributors to soil ecology and play several important roles in soil formation and health:

1. **Burrowing Activity**: As earthworms burrow through fine sediments, they create a network of channels within the soil. This tunneling activity enhances soil aeration, allowing air to penetrate deeper into the soil profile. Improved aeration is essential for the respiration of plant roots and soil microorganisms, promoting a healthy soil ecosystem.

2. **Water Drainage**: The channels created by earthworm burrowing also facilitate increased water seepage. This enhanced drainage helps prevent waterlogging, which can be detrimental to plant growth. By improving the soil's ability to retain and drain water, earthworms contribute to better moisture conditions for plants.

3. **Soil Texture Improvement**: The physical disturbance caused by earthworm activity can lead to improved soil structure. As they move through the soil, earthworms help to mix different soil layers, promoting a more uniform texture that can enhance nutrient availability and root penetration.

4. **Nutrient Enrichment**: Earthworms deposit waste material, known as "castings," along their burrows. These castings are rich in both organic and inorganic materials that the earthworms ingest during their feeding process. The nutrient-rich castings serve as "soil enrichers," providing essential nutrients to plants and enhancing soil fertility. This process is often referred to as "vermicomposting," where earthworm activity contributes to the natural recycling of organic matter.

5. **Nature's Burrows**: Due to their constant activity of turning over the soil and enriching it with their waste materials, earthworms are often referred to as "nature's burrows." Their presence is a strong indicator of healthy soil, as they contribute to the overall biodiversity and functionality of the soil ecosystem.

The Role of Red Ants

In addition to earthworms, other soil-dwelling organisms, such as red ants, also play a significant role in soil formation and health (Fig.3.11 (B)):

1. **Burrowing and Nesting**: Red ants are commonly found in various soil types, from fine to coarse deposits. They create nests that can disturb the soil structure, contributing to the mixing of soil layers. Their burrowing activities can also enhance soil aeration and drainage, similar to the effects of earthworms.

Fig.3.11: Burrowing Earthworms marked 'a' in (A) and Red Ants' marked 'r' in (B) as agents for Biological Weathering

2. **Sediment Removal**: Red ants are actively involved in the removal of fine sediments from the soil to the ground surface. This behavior can help to aerate the soil and promote the breakdown of organic matter, further contributing to soil health. By moving sediments, ants can also influence the distribution of nutrients and organic materials within the soil profile.

3. **Adaptability to Conditions**: Red ants thrive in a range of environmental conditions, from partially wet to dry climates, including arid and semi-arid regions. Their adaptability allows them to play a crucial role in soil formation across diverse ecosystems.

Conclusion

The biochemical activities of animals, such as earthworms and red ants, are essential for soil formation and health. Through their burrowing and nesting behaviors, these organisms enhance soil aeration, improve water drainage, and contribute to nutrient cycling. The interactions between these animals and the soil environment highlight the importance of biological processes in shaping soil characteristics and promoting ecosystem health. Understanding the roles of these organisms in soil formation can inform land management practices and conservation efforts aimed at maintaining healthy and productive soils. The intricate relationships between soil organisms and their environment underscore the dynamic nature of soil ecosystems and their critical role in supporting plant life and overall biodiversity.

The Role of Bioturbation in Soil and Sediment Dynamics

Bioturbation refers to the reworking of soils and sediments by living organisms, including both animals and plants. This process plays a crucial role in shaping soil structure, enhancing nutrient cycling, and facilitating various weathering processes. The activities of bioturbators can significantly alter the physical and chemical properties of the soil and sediment, leading to a range of ecological and geological effects.

Effects of Bioturbation

1. **Changing Texture of Sediments**: Bioturbation can lead to the mixing and redistribution of sedimentary grains, resulting in changes to the texture of the soil or sediment. This mixing can enhance soil structure, improve aeration, and increase the availability of nutrients for plants and microorganisms.

2. **Bio-Irrigation**: The activities of bioturbators can facilitate bio-irrigation, where the movement of organisms through the soil creates channels that allow water to penetrate more deeply. This increased water infiltration can enhance the moisture availability for plant roots and promote chemical weathering processes.

3. **Displacement of Microorganisms and Particles**: As bioturbators move through the soil, they can displace microorganisms and other particles, leading to changes in the microbial community structure and diversity. This displacement can influence nutrient cycling and the overall health of the soil ecosystem.

Common Bioturbators

Various organisms (Fig.3.12 (b)) contribute to bioturbation, each playing a unique role in the reworking of soils and sediments:

Fig.3.12: Pholadidae shell(A), Hideoaut bores in Shale(B) and Land activities by Man in Honduras (C)

- **Annelids**: Ringed worms, such as oligochaetes, are well-known bioturbators. Their burrowing activities create channels in the soil, enhancing aeration and water infiltration. As they ingest soil and organic matter, they contribute to nutrient cycling through their waste products.

- **Bivalves**: Organisms like mussels and clams also engage in bioturbation. They can burrow into sediments, creating structures that alter the sedimentary environment. Their feeding activities can help to recycle nutrients and organic matter within the sediment.

- **Gastropods**: Snails and slugs contribute to bioturbation through their movement and feeding habits. They can help to break down organic matter and facilitate the mixing of soil layers.

- **Holothurians**: Sea cucumbers are important bioturbators in marine environments. They ingest sediment and organic material, processing it through their digestive systems and excreting nutrient-rich waste.

Faunal Activities

The activities of bioturbators encompass a range of behaviors that contribute to soil and sediment dynamics:

- **Burrowing**: Many organisms, such as annelids and bivalves, create burrows that enhance soil structure and aeration. These burrows can also serve as habitats for other organisms.
- **Ingestion and Defecation**: The ingestion of soil grains and organic matter by bioturbators leads to the breakdown of materials and the release of nutrients through their waste products. This process is essential for nutrient cycling in the soil.
- **Construction and Maintenance of Galleries**: Some organisms construct complex burrow systems that can alter the flow of water and nutrients within the soil. These galleries can also provide refuge for other organisms.
- **Infilling of Abandoned Dwellings**: When bioturbators abandon their burrows, these spaces can be infilled with sediment, contributing to the mixing and redistribution of soil materials.
- **Displacement and Mixing of Sedimentary Grains**: The movement of bioturbators through the soil leads to the displacement and mixing of sedimentary grains, which can enhance soil fertility and structure.

Rock Boring and Weathering

Certain organisms, such as members of the family Pholadidae (piddocks), bore into rocks or soils for protection. They can achieve this by either scraping away grains or secreting acids that dissolve rock or soil particles. This behavior not only provides a safe habitat for the organisms but also contributes to the weathering of the rock material, facilitating the breakdown of solid rock into smaller particles that can be further weathered and transformed into soil.

Conclusion

Bioturbation is a vital process that significantly influences soil and sediment dynamics. The activities of various organisms, from earthworms to bivalves and gastropods, contribute to the reworking of soils, enhancing aeration, nutrient cycling, and moisture retention. By facilitating these processes, bioturbators play a crucial role in soil formation and health, ultimately supporting plant growth and ecosystem stability. Understanding the mechanisms and impacts of bioturbation is essential for effective land management and conservation practices, as it highlights the interconnectedness of biological and geological processes in shaping the Earth's surface.

Piddocks and Their Role in Bioturbation

Piddocks, a unique group of bivalve mollusks, possess shells that do not fully enclose their bodies, allowing them to burrow into various substrates, including wood and rock. These organisms rely on their burrowing capabilities for protection, often using other materials, such as shells, to shield themselves from predators and environmental conditions. Piddocks are most commonly found in low intertidal zones, typically at or just below the low tide line, where they can burrow to depths of approximately 1 meter in soft sand or muddy substrates. Their burrowing activities contribute to bioturbation, which plays a significant role in sediment dynamics and the overall health of coastal ecosystems.

Human Activities and Environmental Degradation

Human activities have a profound impact on weathering processes and can lead to environmental degradation and desertification. Some of the key activities contributing to these negative changes include:

1. **Clearing of Vegetation**: The use of fire and other methods to clear vegetation can lead to soil erosion and loss of soil structure. Vegetation plays a crucial role in stabilizing soil and maintaining moisture levels, and its removal can expose soil to increased weathering and degradation.

2. **Poor Land Cultivation Practices**: Unsustainable agricultural practices, such as overgrazing, monoculture, and improper tillage, can degrade soil health and lead to desertification. These practices can disrupt soil structure, reduce organic matter, and increase erosion.

3. **Excavation of Mine Shafts**: Mining activities can significantly alter landscapes and disrupt natural processes. The excavation of mine shafts can

lead to soil compaction, loss of vegetation, and contamination of soil and water resources.

4. **Environmental Pollution**: Pollution from industrial activities, agricultural runoff, and urban development can change soil chemistry and the composition of atmospheric gases. This pollution can lead to acidification of soils and water bodies, negatively impacting plant and animal life.

These human-induced changes contribute to environmental degradation and can exacerbate global issues such as climate change and global warming. The alteration of natural processes can slow down the positive gains achieved since the dawn of industrialization, leading to a cycle of environmental decline.

Microbial Activity in Weathering

Microorganisms play a crucial role in weathering processes, particularly through their ability to break down rock minerals and alter chmical compositions (Fig.3.13).. One notable example of microbial activity in weathering is the role of lichens:

Fig.3.13: Rock covered wirh Lichen, L.Superior, USA

- **Lichens**: Lichens are symbiotic associations between fungi and algae. They often establish themselves on barren rocks, where they can initiate the weathering process. The fungi release organic acids and other chemicals that break down rock minerals, creating weak zones in the rock. The minerals released during this process are then consumed by the algae, which benefit from the nutrients provided by the fungi.

- **Chemical Weathering**: As lichens continue to grow and metabolize, they create holes and gaps in the rock, further exposing it to physical and chemical weathering. The metabolic processes of microorganisms also release carbon dioxide, which can combine with water to form carbonic acid. This acid plays a significant role in chemical weathering, facilitating the dissolution of minerals and contributing to soil formation.

Organic Material and Soil Fertility

Organic material generated from the decomposition of plants and animals is essential for soil health and fertility. When the metabolism of these organisms ceases, their remains contribute to the formation of humus, a vital component of healthy soil:

- **Nutrient Provision**: The decomposition of organic matter enriches the soil with nutrients, which are essential for the growth of plants and other organisms. Humus improves soil fertility by providing a reservoir of nutrients that can be readily available to plants.
- **Water and Air Retention**: Humus increases the soil's capacity to retain water and air, both of which are critical for plant growth. Well-structured soil with adequate organic matter can hold moisture more effectively, reducing the need for frequent irrigation and enhancing drought resilience.
- **Continuity of Weathering Processes**: The presence of organic material in the soil contributes to ongoing weathering processes. As organic matter decomposes, it releases organic acids that can further enhance chemical weathering, promoting the breakdown of minerals and the formation of new soil.

Conclusion

The interplay between bioturbation, microbial activity, and organic material is essential for soil formation and health. Piddocks and other bioturbators contribute to sediment dynamics, while microorganisms, such as lichens, facilitate weathering through their metabolic activities. Additionally, the decomposition of organic matter enriches the soil, enhancing its fertility and capacity to support plant life. However, human activities pose significant challenges to these natural processes, leading to environmental degradation and desertification. Understanding these interactions is crucial for developing sustainable land management practices and mitigating the impacts of human activities on soil and ecosystem health.

1.5 Factors Affecting Weathering Processes

The type of weathering that affects rock mass, rate and extent of weathering achieved depend upon several factors such as;

(a) ***Climate***: The climate dictates the type of weathering processes that operates in any region based on the amount of water available and the temperature at which the processes occur. Usually, chemical reactions are enhanced and faster at higher temperatures while frost wedging is prevalent in colder climates.

(b) ***Topography***: The slope angle for any terrain determines the energy of the weathering system by controlling the rate at which water seeps through the rock mass down-slope. Generally, higher or tectonically active areas with steeper slopes have more dynamic weathering systems than flat plains.

(c) ***Rock Type***: Rock type determines resistance that a rock offers to weakening agents that operate in a particular environment. Each rock type is composed of a particular set of minerals, which are joined together by chemical bonding or cementing processes that occurred in the geologic past. In general, siliciclastic rocks offer a higher resistance to chemical weathering compared to carbonates. Bonding is such rocks is stronger compared to bonding for non-siliciclastic rocks.

(d) ***Rock Structure***: Jointed or faulted rocks contain many planes of weakness along which weathering agents such as water penetrate into the rock mass to contribute to chemical weathering (Fig.3.14 (A)). In the tropics and sub-tropics, intense weathering that occurs under the hot and humid conditions produces thick profiles, which may be up to 100 meters or more.

Fig.3.14: Weathering patterns in Granite(A) and in Quartz Monzonite(B)

Weathering processes penetrate down planes of weakness referred to as discontinuities (Fig.3.14 (B)). Such planes of weakness include faults and joints in the rockmass. The weathering process then attacks faces of the joint-bounded blocks, penetrating into the solid blocks. Weathering preferentially attacks the corners and edges of the joint blocks, causing them to become rounded. This action is assisted by stress release, which causes flaking into curved shells as an exfoliation process which occurs prevalently to form rounded corestones

(e) ***Erosion***: Erosion being the removal of weakened rock material from *in-situ*, is undertaken primary by gravity as an agent of transportation. However, gravity acts in concert with running water. The dynamism and efficiency of erosion determines the rate of removal of weathered material thereby exposing fresh rock mass to further weathering processes while preserving deeply weathered profiles.

(f) ***Time***: This is the duration taken by the same type of weathering process uninterrupted by climatic change, earth movements, and other factors. It determines the degree and depth to which rocks are weakened.

Typical weathering categories as a guide to textural changes that are likely to be encountered in a deteriorating rock mass is as shown in table 3.1.

Table 3.1: Weathering grade classification

Weathering grade	Description
Fresh	No visible sign of weathering. Rock fresh and crystals still bright.
Slightly weathered	Discontinuities are strained or discolored and may contain a thin filling of altered material. Discoloration may extend into the rockfrom discontinuities to a distance of 20% of discontinuity spacing.
Moderately weathered	Slight discoloration extends from discontinuity planes for a distance of more than 20%. Discontinuities may contain a thin filling of altered material. Partial opening of grain boundaries is observed.
Highly weathered	Discoloration extends throughout the rock and the rock materialis partially friable. The original texture of the rock is preserved but grains are already separated.
Completely weathered	The rock is completely discolored, decomposed and in friable conditions. The external appearance is that of a soil. Internally, the rock texture is partially preserved but the grains are completely separated.
Residual soils	Not included

Rock structure and joint patterns significantly influence the mechanical weathering of bedded rocks, as illustrated in Figure 3.15. The orientation, spacing, and characteristics of joints—natural fractures in the rock—play a crucial role in determining how these rocks respond to weathering processes.

In scenario 'A', the failure of the rock can lead to the formation of large rock slabs. These slabs may detach from the parent rock due to the expansion and contraction caused by temperature fluctuations, freeze-thaw cycles, or other environmental factors. The presence of well-defined joints allows for the easy separation of these slabs, which can then slide or fall away from the rock face, contributing to the overall erosion of the landscape.

In contrast, scenario ‘B’ illustrates the formation of small tabular blocks that are particularly susceptible to toppling due to gravitational forces. These blocks typically arise in environments where the joints in the rock are closely spaced or when the rock layers are thin and horizontally oriented. The closely spaced joints create natural planes of weakness, allowing for the easier separation of these blocks from the surrounding rock mass. As weathering processes, such as freeze-thaw cycles, thermal expansion, and chemical weathering, gradually weaken the bonds between these blocks and the underlying rock, they can become dislodged. Once these blocks lose their structural integrity, they may topple over, contributing to the mechanical breakdown of the rock mass and altering the landscape.

The susceptibility of these small tabular blocks to toppling is particularly significant in areas with steep slopes or where gravitational forces are pronounced. As these blocks fall or slide, they can create additional hazards, such as rockfalls, which pose risks to infrastructure, roadways, and human safety. The dynamic nature of this process underscores the importance of understanding how mechanical weathering operates in different geological contexts.

The differences in failure modes between scenarios ‘A’ and ‘B’ highlight the critical role that rock structure and joint patterns play in influencing the rate and style of mechanical weathering. In scenario ‘A’, larger rock slabs may detach and slide down slopes, while in scenario ‘B’, smaller blocks may topple, leading to different erosion patterns and landscape features. This variability in failure mechanisms is essential for geologists and engineers to consider, as it can significantly impact slope stability, rockfall hazards, and the overall integrity of geological formations.

Understanding these processes is vital for effective land management and construction practices. For instance, in areas prone to landslides or rockfalls, engineers can design structures that mitigate these risks by taking into account the specific joint patterns and rock structures present. Additionally, geologists can use this knowledge to assess the stability of slopes and predict potential failure events, thereby informing land-use planning and development strategies.

By studying the relationship between rock structure and weathering, we can better predict how bedded rocks will behave under various environmental conditions.

This predictive capability is crucial for developing effective strategies to manage natural resources, protect infrastructure, and ensure the safety of communities living in proximity to potentially unstable geological formations. Ultimately, a comprehensive understanding of these processes not only enhances our knowledge of geological systems but also informs practical applications in engineering, environmental management, and disaster preparedness.

Fig.3.15: Mechanical rock breakage caused to different Sandstone Beds

CHAPTER TWO

TRANSPORTATION, EROSION AND DIAGENESIS

Weathering processes generate a variety of materials, including sediments and rock fragments, which can be displaced to different localities by various transporting agents. These agents play a crucial role in the erosion, transportation, and deposition of weathered materials, shaping landscapes over time. The primary transporting agents include running water, moving ice (glaciers), wind, and gravity (Fig.3.16).

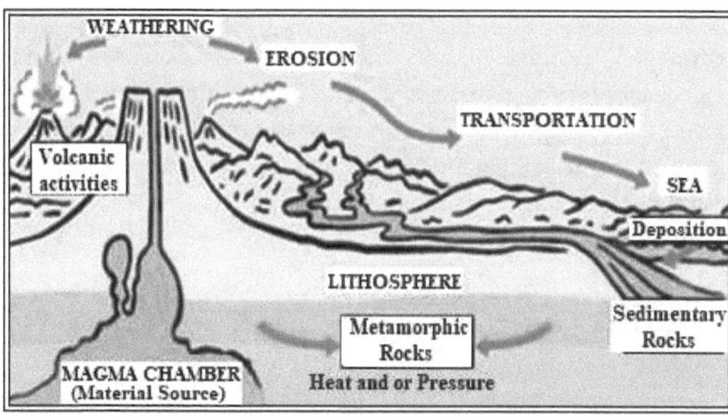

Fig.3.16: General processes of a Sedimentary Cycle

2.1. Running Water

Running water is one of the most significant agents of erosion and transportation. It can carry sediments from their source to new locations through several processes:

- **Erosion of Stream Channels**: As water flows over the land, it erodes the stream channels, breaking down rocks and sediments along the banks and bed. This erosion can lead to the formation of valleys and canyons over time. The sediments that are eroded from the stream channels are then transported downstream, where they may be deposited in riverbeds, floodplains, or deltas.

- **Sediment Transport**: The ability of running water to transport sediments depends on the flow velocity and the size of the particles. Larger particles, such as gravel and cobbles, require faster water flow to be moved, while smaller particles, like silt and clay, can be transported even in slower-moving water. The continuous movement of water helps to sort sediments by size, leading to distinct sedimentary layers in depositional environments.

2.2. Moving Ice (Glaciers)

Glaciers are powerful agents of erosion and transportation, particularly in cold regions:

- **Glacial Erosion**: As glaciers advance, they scrape and grind the underlying rock, producing a wide range of sediments, from fine silt to large boulders. This process is known as glacial erosion and can significantly alter the landscape, creating features such as U-shaped valleys and fjords.

- **Sediment Transport**: Glaciers can transport vast amounts of material over long distances. As glaciers melt, they release the sediments they have carried, depositing them as till or outwash in various landforms, such as moraines and glacial plains. The sediments deposited by glaciers are often unsorted, containing a mix of particle sizes.

3. Wind

Wind is a significant transporting agent in arid and semi-arid regions, where vegetation is sparse:

- **Transportation of Fine Sediments**: Wind can transport fine sediments, such as sand and silt, over considerable distances. This process is particularly evident in desert environments, where strong winds can lift and carry particles, leading to the formation of features like dunes and loess deposits.

- **Erosion by Wind**: Wind can also cause erosion by removing loose particles from the surface of rocks and soils. This process, known as deflation, can lead to the formation of depressions and the exposure of underlying materials.

4. Gravity

Gravity is a fundamental force that influences the movement of materials on slopes:

- **Mass Wasting**: Gravity causes materials to move downslope through processes such as landslides, rockfalls, and soil creep. These mass wasting events can transport weathered materials from higher elevations to lower areas, contributing to the reshaping of landscapes.
- **Sediment Accumulation**: As materials are transported downslope, they can accumulate in valleys, on hillsides, or at the base of slopes, forming deposits that can later be reworked by other transporting agents.

Conclusion

The transportation of weathered materials by agents such as running water, moving ice, wind, and gravity plays a crucial role in shaping the Earth's surface. Over millions of years, these processes have continuously broken down and moved rock materials, leading to the formation of diverse landforms and sedimentary deposits. Understanding the dynamics of these transporting agents is essential for comprehending the broader processes of erosion, sedimentation, and landscape evolution. The interplay between weathering and transportation highlights the interconnectedness of geological processes and their impact on the environment.

These small pieces, if hosted by stream systems, get washed further downstream where they settle to the bottom of the rivers, lakes, and oceans. The deposited material may be reworked by erosional agents and transported again to a different site. The removal of existing material (erosion) by agents such as running water is favored by the weight of the sediments being transported. Ultimately, the material ends up being layered into a final depositional environment. The layers get pressed down due to self-weight of the sediments as the bottom layers slowly turn into bedrocks successively upwards.

Diagenesis: The term diagenesis is used to describe all the chemical, physical, and biological changes, including cementation, that are subjected to sediment after their initial deposition, exclusive of surface weathering. Some of the processes cause the sediment to consolidate forming a compact solid material. Young sedimentary rocks, especially those of Quaternary age (the most recent period of the geologic time scale) are often still unconsolidated. As sediment transportation builds up, the overburden (lithostatic) pressure rises and a process known as lithification takes place. Sedimentary rocks are often saturated with seawater or groundwater, in which minerals can dissolve or from which minerals can precipitate.

Thus, **Diagenesis** is a comprehensive term that describes the suite of chemical, physical, and biological changes that sediments undergo after their initial deposition, excluding surface weathering processes. This intricate series of transformations is essential for converting loose sediments into solid rock, and it encompasses a variety of mechanisms, including cementation, compaction, and mineral alteration. Understanding diagenesis is crucial for geologists, as it provides insights into the history and evolution of sedimentary rocks, their resource potential, and their behavior under various geological conditions.

Processes of Diagenesis

1. **Cementation**: Cementation is a key process in diagenesis where minerals precipitate from groundwater and fill the pore spaces between sediment grains. Common cementing agents include silica, calcite, and iron oxides. As these minerals crystallize, they bind the sediment particles together, transforming loose grains into a coherent rock mass. The type and amount of cement can significantly influence the rock's porosity and permeability, affecting its ability to store and transmit fluids. For instance, a rock with abundant calcite cement may be less porous than one with silica cement, which can have implications for groundwater flow and hydrocarbon reservoirs.

2. **Compaction**: Compaction occurs as sediments are buried under additional layers of material, leading to an increase in lithostatic pressure. This pressure causes sediment grains to be squeezed closer together, reducing pore space and expelling interstitial water. In clay-rich sediments, compaction can lead to the expulsion of up to 60% of the water contained in the pore spaces. The mechanical reorganization of grains during compaction can also enhance the rock's strength and stability. Compaction is particularly important in the formation of shales, where fine-grained sediments are subjected to significant pressure over time.

3. **Mineral Alteration**: During diagenesis, the original mineral composition of sediments can change due to various chemical reactions. For example, feldspar may alter to clay minerals through hydrolysis, while calcite can precipitate from seawater or groundwater. These alterations can affect the rock's physical and chemical properties, influencing its durability and reactivity. Additionally, the dissolution of certain minerals can lead to increased pore space, while the precipitation of new minerals can fill these spaces, further complicating the rock's structure.

4. **Leaching**: Leaching is a process where certain minerals dissolve in water and are removed from the sediment, increasing pore space and altering the rock's composition. This process can occur in environments with high rainfall or groundwater flow, where acidic conditions promote the dissolution of soluble minerals. Leaching can significantly impact the geochemistry of sedimentary rocks, affecting their mineralogy and potential resource value.

5. **Biochemical Processes**: Biological activity plays a significant role in diagenesis. Microorganisms, such as bacteria and fungi, can influence the mineral composition of sediments through their metabolic processes. For instance, certain bacteria can precipitate minerals like calcite or pyrite, while others may facilitate the breakdown of organic matter, contributing to the formation of hydrocarbons. The presence of organic material can also affect the chemical environment, promoting specific diagenetic pathways.

6. **Temperature and Pressure Effects**: As sediments are buried deeper, the increase in temperature and pressure can stimulate various chemical reactions. Over geological time, these conditions can lead to the transformation of organic material into coal or hydrocarbons. For example, the gradual increase in temperature and pressure can convert peat into lignite, then into bituminous coal, and eventually into anthracite. This transformation is a critical aspect of the carbon cycle and has significant implications for energy resources.

Transition to Metamorphism

As diagenesis progresses, the conditions of temperature and pressure may eventually exceed those typical of diagenetic processes, leading to the onset of **metamorphism**. This transition marks a significant change in the rock's characteristics, as the minerals undergo further alteration due to increased heat and pressure. Metamorphic processes can result in the formation of new minerals, changes in texture, and the development of foliation or other structural features. Understanding the boundary between diagenesis and metamorphism is essential for geologists, as it helps delineate the conditions under which different rock types form and evolve.

Implications of Diagenesis

The processes of diagenesis have profound implications for various fields, including geology, petroleum engineering, and environmental science. For instance:

- **Resource Exploration**: Knowledge of diagenetic processes is crucial for identifying and evaluating potential hydrocarbon reservoirs. The porosity and permeability of sedimentary rocks, which are influenced by diagenesis, determine their capacity to store and transmit oil and gas.

- **Groundwater Management**: Understanding how diagenesis affects the porosity and permeability of aquifers is essential for effective groundwater management. Changes in mineral composition and pore space can influence the availability and quality of groundwater resources.

- **Geotechnical Engineering**: In construction and civil engineering, understanding diagenesis is vital for assessing the stability and integrity of sedimentary rock formations. Knowledge of how diagenetic processes affect rock properties can inform decisions related to site selection, foundation design, and slope stability analysis. For example, engineers must consider the potential for compaction and cementation when designing structures on sedimentary rocks to ensure they can withstand the loads imposed by buildings and infrastructure.

- **Environmental Impact Assessments**: Diagenesis can influence the behavior of contaminants in sedimentary environments. Understanding how sediments change over time can help predict how pollutants may migrate through groundwater systems or be retained in sediments. This knowledge is crucial for developing effective remediation strategies for contaminated sites.

- **Paleoclimate Studies**: Diagenetic processes can also provide insights into past environmental conditions. The mineralogical and chemical changes that occur during diagenesis can reflect the depositional environment and the conditions under which sediments were formed. By studying these changes, geologists can reconstruct ancient climates and understand how they have evolved over geological time.

Precipitating minerals reduce pore space in sedimentary rocks and expel water incorporated in the sediments, leading to cementation, which binds fragments together and increases the rock's compactness and resistance to deformation. As sedimentation continues, older rock layers become buried deeper, increasing lithostatic pressure and contributing to compaction, where grains are mechanically reorganized, often resulting in the expulsion of up to 60% of interstitial water in clay. Compaction can also create solution pressure, leading to the dissolution of some grains, while leaching can increase pore space by dissolving minerals without re-precipitation. Biochemical processes, such as the activity of bacteria, fungi, and plants, also influence diagenetic changes. Additionally, the burial of rocks due to sedimentation raises pressure and temperature, stimulating chemical reactions that can transform organic material into coal over geological time. Ultimately, as conditions intensify, diagenesis transitions into metamorphism, leading to the formation of metamorphic rocks.

Conclusion

In summary, diagenesis is a complex and multifaceted process that plays a critical role in the transformation of sediments into sedimentary rocks. It encompasses a range of physical, chemical, and biological changes that not only affect the properties of the rocks but also influence their behavior in geological settings. Understanding diagenesis is essential for geologists and engineers alike, as it provides valuable insights into the history of sedimentary formations, their potential resources, and their responses to environmental changes over time.

By studying diagenetic processes, we can better predict how sedimentary rocks will behave under various conditions, ultimately aiding in effective land management, resource exploration, and environmental protection. The intricate interplay of factors involved in diagenesis highlights the dynamic nature of geological processes and underscores the importance of continued research in this field to enhance our understanding of Earth's geological history and its ongoing evolution.

CHAPTER THREE

CLASSIFICATION OF SEDIMENTARY ROCKS

Sedimentary rocks can be classified using various schemes, with two primary approaches focusing on mineralogical composition and genetic (or process-based) considerations. This classification is essential for understanding the origins, characteristics, and uses of different sedimentary rock types. Below is a detailed overview of both compositional and genetic classification schemes.

Compositional Classification Scheme

The compositional classification scheme is based on the mineralogical content of sedimentary rocks and results in the following seven groups:

1. **Siliciclastic Sedimentary Rocks**:
    - **Definition**: These rocks are predominantly composed of silicate minerals, primarily derived from the weathering and erosion of pre-existing rocks.
    - **Formation**: The sediments that constitute siliciclastic rocks are transported as suspended loads in water or by sediment gravity flows, such as debris flows and turbidity currents.
    - **Subdivisions**:
        - **Conglomerates and Breccias**: Composed of larger clasts (>2 mm) that are cemented together. Conglomerates have rounded clasts, while breccias consist of angular clasts.
        - **Sandstones**: Composed of sand-sized particles (0.0625 to 2 mm) and can be further classified based on their mineral content (e.g., quartz sandstone, feldspathic sandstone).
        - **Mudrocks**: Composed of silt and clay-sized particles (<0.0625 mm), including siltstones, mudstones, and shales.

2. **Carbonate Sedimentary Rocks**:

- **Definition**: These rocks are primarily composed of carbonate minerals, which are formed from the precipitation of calcium carbonate and other carbonate minerals.
- **Common Minerals**:
  - **Calcite** (rhombohedral $CaCO_3$)
  - **Aragonite** (orthorhombic $CaCO_3$)
  - **Dolomite** ($CaMg(CO_3)_2$)
- **Examples**: Limestone (primarily composed of calcite) and dolostone (rich in dolomite) are common carbonate rocks, often formed in marine environments.

3. **Evaporite Sedimentary Rocks**:

- **Definition**: These rocks are formed from the crystallization of minerals from supersaturated waters, typically in arid environments where evaporation rates are high.
- **Common Minerals**:
  - **Carbonates** (e.g., calcite)
  - **Chlorides** (e.g., halite)
  - **Sulfates** (e.g., gypsum)
- **Examples**: Bedded evaporite rocks rich in halite (rock salt), gypsum, and anhydrite are typical representatives of this group.

4. **Organic-Rich Sedimentary Rocks**:

- **Definition**: These rocks contain significant amounts of organic material, generally exceeding 3% total organic carbon.
- **Examples**: Coal (formed from plant material) and oil shale (rich in kerogen) are key examples, serving as important source rocks for oil and natural gas.

5. **Siliceous Sedimentary Rocks**:
    - **Definition**: Composed almost entirely of cryptocrystalline silica ($SiO2$), these rocks are formed from the accumulation of siliceous organisms or the precipitation of silica from solution.
    - **Examples**: Chert, flint, and diatomite are common siliceous rocks, often formed in marine environments or from the remains of diatoms.
6. **Iron-Rich Sedimentary Rocks**:
    - **Definition**: These rocks contain iron content exceeding 15%.
    - **Common Forms**: Banded iron formations (BIFs) and ironstones are typical examples, often formed in ancient marine environments where iron was precipitated from seawater.
7. **Phosphatic Sedimentary Rocks**:
    - **Definition**: Composed of phosphate minerals with over 6.5% phosphorus.
    - **Examples**: Deposits of phosphate nodules, bone beds, and phosphatic mudrocks are significant for their role in agriculture and as a source of phosphorus.

Genetic Classification Scheme

The genetic classification scheme categorizes sedimentary rocks based on the processes responsible for their formation. This approach divides sedimentary rocks into four main groups:

1. **Clastic Sedimentary Rocks**:
    - **Definition**: Formed from the accumulation and lithification of clastic sediments derived from the weathering of pre-existing rocks.
    - **Examples**: Sandstones, mudrocks, conglomerates, and breccias fall under this category.
2. **Chemical Sedimentary Rocks**:
    - **Definition**: Formed from the precipitation of minerals from solution, often in evaporative environments.

- **Examples**: Evaporites (e.g., rock salt, gypsum) and some limestones that form through chemical processes.

3. **Biochemical (or Biogenic) Sedimentary Rocks**:

   - **Definition**: Formed from the accumulation of biological materials, such as shells, plant debris, and other organic matter.

   - **Examples**: Limestones formed from the accumulation of shell fragments and coal formed from plant material.

4. **Miscellaneous ("Other") Sedimentary Rocks**:

- **Definition**: The category of miscellaneous or "other" sedimentary rocks encompasses those formed through processes that do not fit neatly into the traditional classifications of clastic, chemical, or biochemical sedimentary rocks. This includes rocks resulting from unique geological events such as impacts from extraterrestrial bodies, volcanic activity, and other minor processes that lead to sedimentary formations. These rocks often exhibit characteristics that reflect their diverse origins and the complex interactions of geological forces.

- **Examples**:

  - **Pyroclastic Rocks**: While primarily classified as igneous due to their volcanic origin, pyroclastic rocks can also be treated as sedimentary rocks because they undergo transport and deposition by sedimentary agents, such as wind and water. Pyroclastic materials, which include ash, pumice, and volcanic fragments, can be ejected during explosive volcanic eruptions and subsequently settle out of the atmosphere or be carried by water to form sedimentary deposits. When these materials accumulate and lithify, they can create sedimentary rock formations that reflect the dynamic processes of volcanic activity.

  - **Impact Breccias**: Formed from the shattering of rocks due to high-energy impacts, such as meteorite strikes, impact breccias can also be classified within this category. The intense pressure and heat generated during an impact can break apart surrounding rock, leading to the formation of angular fragments that are then cemented together. These rocks provide valuable information about the history of impacts on Earth and other planetary bodies.

- **Volcaniclastic Rocks**: These rocks are formed from the accumulation of volcanic materials, such as ash and tephra, that are transported by wind or water. Volcaniclastic deposits can exhibit characteristics of both sedimentary and volcanic rocks, depending on their composition and the processes involved in their formation.

Conclusion

The classification of sedimentary rocks is essential for understanding their origins, characteristics, and the processes that shape them. By employing both compositional and genetic classification schemes, geologists can gain valuable insights into the geological history of an area, the environmental conditions present at the time of deposition, and the potential resources these rocks may contain. Compositional classification focuses on the mineral content and texture of the rocks, while genetic classification considers the processes of formation, such as clastic, chemical, and biochemical origins. This dual approach allows for a more nuanced understanding of sedimentary rocks and their formation.

A comprehensive approach to classification not only aids in academic study but also has significant practical implications in various fields. For instance, in resource exploration, understanding the types of sedimentary rocks present can help geologists identify potential reservoirs for oil, natural gas, and groundwater. In environmental geology, knowledge of sedimentary rock types can inform assessments of soil stability, erosion potential, and contamination risks. Additionally, in construction, recognizing the properties of different sedimentary rocks is crucial for determining their suitability as building materials and for ensuring the stability of structures built on or with these materials.

Understanding the diverse categories of sedimentary rocks, including the miscellaneous group, allows geologists to appreciate the complexity of geological processes and the dynamic nature of the Earth's crust. This complexity reflects the interplay of various factors, such as sediment source, transport mechanisms, depositional environments, and post-depositional alterations. By studying these categories, geologists can better interpret the geological history of a region, including past climate conditions, tectonic activity, and biological evolution.

This knowledge is crucial for effective resource management, as it enables geologists to make informed decisions regarding the sustainable extraction and use of natural resources. Furthermore, it plays a vital role in environmental assessment, helping to identify potential hazards and mitigate risks associated with geological processes. Ultimately, the advancement of geological sciences relies on a thorough understanding of sedimentary rock classification, as it provides the foundation for ongoing research and exploration in the field. By integrating classification with modern technologies and interdisciplinary approaches, geologists can continue to unravel the complexities of the Earth's geological history and its implications for the future.

CHAPTER FOUR

CHARACTERISTICS OF SILICICLASTIC ROCKS

Siliciclastic rocks consist mainly of silica in chemical composition. Mineralogy of a clastic rock generally depends on the type of material that is supplied from the source area, the manner of transport to the place of deposition and the stability of a particular mineral. The stability of the major rock forming minerals (their resistance to weathering) is expressed by Bowen's reaction series. In this series, quartz is most stable, followed by feldspar, micas, and other less stable minerals that are only present when little weathering has occurred. Generally, siliciclastic rocks evolved mainly as a result of transportation i.e. Clastic sedimentary rocks. They are dominated by free silica i.e. α-quartz and other silicate minerals particularly feldspar, rock (lithic) fragments, clay minerals, and mica. Other minerals if present as accessories are of less significance. The siliciclastic rocks are subdivided based on diameter of dominant fragments (Udden- Wentworth grain size scale). The scale divides the rocks into three classes (Table 3.1). Gravel size material is the coarsest while mud size material is the finest.

Table 3.1: Grain size classification for common siliciclastics

Grain size (Diameter)		**Sediment Type**	**Rock Name**
Diam. > 2mm	Gravel	Rounded fragments	Conglomerate
		Angular fragments	Breccia
$\frac{1}{16}$ < diam. < 2 mm	Sand		Sandstone
$\frac{1}{256}$ < diam, <$\frac{1}{16}$ mm	Mud		Siltstone
diam.<$\frac{1}{256}$ mm			Shale

(a) Conglomerate and breccia: Gravels represented by conglomerate and breccia are differentiated based on shape of the clasts as noted from the table.

Conglomerate: A rock consisting of individual clasts within a finer-grained matrix that are cemented together. Conglomerates consist of rounded fragments. They can be differentiated from breccia,which consists of angular clasts. The conglomerates can be grouped into two groups i.e. Paraconglomerates and Orthoconglomerates based on the amount of smaller grains present. Paraconglomerates consist of a matrix-supported rock that contains at least 15% sand-sized or smaller grains (<2 mm), the rest being larger grains of varying sizes. Orthoconglomerates consist of a clast-supported rock with less than 15% matrix of sand and finer particles. Conglomerates are also classified based onthe dominant size of the clasts. The sizes are of the form;

- Granule conglomerate; size range 2–4 mm
- Pebble conglomerate; size range 4–64 mm
- Cobble conglomerate; size range 64–256 mm
- Boulder conglomerate; size >256 mm

A spectacular example of conglomerate occurs at Montserrat, near Barcelona. Here erosion has created vertical channels giving the characteristic jagged shapes for which the mountain is named (Montserrat literally means "jagged mountain"). The rock is strong and is used as a suitable building material. Indeed the monastery buildings shown in the illustration were constructed using the conglomerate. Rounded to sub-rounded clasts marked 'c' are as shown in the inset (Fig.3.17).

Fig.3.17: Geomorphological view of Conglomerate (A) and surface exposure - inset(B), montserrat, Barcelona Spain, June 2012

***Breccia*:** Breccia refers to a type of rock that is composed of broken fragments of minerals or rock material that are cemented together by a fine-grained matrix. This matrix can be similar to or different from the composition of the fragments themselves (see Fig. 3.18 (A)). Breccias can have a variety of origins, and they are classified into several types based on their formation processes. The typical types of breccia include sedimentary breccia, tectonic breccia, igneous breccia, impact breccia, and hydrothermal breccia.

Types of Breccia

1. **Sedimentary Breccia**:

    - Sedimentary breccia is a clastic sedimentary rock made up of randomly oriented angular to sub-angular clasts of other sedimentary rocks.
    - These breccias are typically formed by processes such as submarine debris flows, avalanches, mud flows, or mass flows in an aqueous medium. The angularity of the clasts indicates that they have not been transported far from their source, preserving their sharp edges.

2. **Igneous Breccia**:

    - Igneous breccia consists of broken fragments of both lava and pyroclastic materials, as well as fragments of intrusive rock.
    - This type of breccia can form during explosive volcanic eruptions, where the force of the eruption shatters the surrounding rock and ejects it into the air, where it subsequently falls back to the ground and consolidates.

3. **Tectonic Breccia**:

    - Tectonic breccia forms in regions where relative movements between two tectonic plates result in the grinding action of two fault blocks as they slide past each other.
    - The intense pressure and friction can break the rocks into fragments, which are then cemented together by mineral growth introduced by groundwater. This type of breccia is often found in fault zones and can provide insights into tectonic activity in the area.

4. **Impact Breccia**:

    - Impact breccia is formed as a result of high-energy impacts, such as meteorite strikes.
    - The extreme pressure and temperature generated by the impact can shatter the surrounding rock, creating a mixture of angular fragments that are cemented together.

5. **Hydrothermal Breccia**:

- Hydrothermal breccia forms when hot, mineral-rich fluids circulate through fractures in the rock, causing the rock to break apart and become cemented by the minerals precipitated from the fluid.
- This type of breccia is often associated with mineral deposits and can indicate areas of past hydrothermal activity.

Example: Miocene San Onofre Breccia

Figure 3.18 illustrates a sedimentary breccia sample (a) and the Miocene San Onofre Breccia cliffs in California. The angular clasts, cemented in the matrix, are marked as 'c' in the inset. Notably, the clasts of pebble and cobble sizes are well cemented and remain intact as exposed on the cliff face. This example highlights the characteristics of sedimentary breccia, showcasing the angularity of the clasts and the effectiveness of the cementing matrix in preserving the structure of the rock.

Conclusion

Breccia is a versatile rock type that provides valuable information about geological processes and environments. Its various forms, each with distinct origins, reflect the dynamic nature of the Earth's crust and the forces that shape it. Understanding breccia and its characteristics can offer insights into sedimentary processes, tectonic activity, and the history of volcanic eruptions, making it an important subject of study in geology.

Fig.3.18: Geomorpholical view of Breccia Cliff (A) and Rock sample inset(B), San Onofre, California December 2012

(b) Sandstones: Sandstones are the most widespread of the siliciclastic rocks, occurring as well-bedded rocks that have been cemented with secondary quartz during diagenesis. The sediments that formed the sandstones were derived mostly from granitic intrusives and granitic gneisses.

Classification of Sandstones

The classification of sandstone rocks varies widely, but most geologists have adopted the Dott scheme, which uses the relative abundance of quartz, feldspar, and lithic framework grains and the abundance of muddy matrix between these larger grains.

Composition of Framework Grains

The relative abundance of sand-sized framework grains determines the first descriptive word in a sandstone name. For naming purposes, the abundance of framework grains is normalized to quartz, feldspar, and lithic fragments derived from other rocks. These are the three most abundant components of sandstones. Other minerals are considered to be accessories and are not used in naming sandstones, even if abundant.

There are three types of sandstones based on the relative abundance of the three components:

- **Quartz Sandstones**: Quartz grains greater than 90%
- **Feldspathic Sandstones**: Quartz grains less than 90% and feldspar grains more than lithic grains
- **Lithic Sandstones**: Quartz grains less than 90% and lithic grains more than feldspar grains

Abundance of Muddy Matrix between Sand Grains

When sand-sized particles are deposited, the space between sand grains either remains open or is filled with mud, i.e., silt and/or clay-sized particles. "Clean" sandstones with open pore space (that may later be filled with cement) are called arenites, while muddy sandstones with abundant muddy matrix (>10%) are known as wackes.

Thus, six sandstone names are possible using descriptors for grain composition (quartz-, feldspathic-, and lithic-) and amount of matrix (wacke or arenite). For example:

- A quartz arenite would be composed of mostly quartz grains (>90%) and have little/no clayey matrix between the grains.

- A lithic wacke would have abundant lithic grains (quartz <90%, remainder would have more lithics than feldspar) and abundant muddy matrix.

Quartz sandstones and feldspathic sandstones are generally referred to as quartzite and arkoses, respectively, while a lithic sandstone that contains muddy matrix (>10%) is referred to as greywacke. Most greywacke beds, being calcic, are fairly stable and form suitable construction materials.

c) Mudrocks

Mudrocks are sedimentary rocks that are composed of at least 50% silt- and clay-sized particles. These relatively fine-grained particles are commonly transported in suspension by turbulent flow in water or air. They become deposited from the suspension when the flow calms down.

Mudrocks are divided into:

- **Siltstones**: Dominated by silt-sized particles

- **Mudstones**: Sub-equal mixture of silt- and clay-sized particles

- **Claystones**: Dominated by clay-sized particles

Mudrock that is fissile is known as shale, regardless of particle size and shape.

Characteristics of Mudrocks

Mudrocks exhibit several key characteristics that distinguish them from coarser sedimentary rocks:

1. **Grain Size**: As mentioned, mudrocks are primarily composed of silt and clay-sized particles, which are less than 0.063 mm in diameter. This fine grain size contributes to their unique physical properties and behavior during deposition and lithification.

2. **Fissility**: One of the defining features of mudrocks, particularly shale, is their ability to split into thin layers or sheets along planes of weakness. This property, known as fissility, is a result of the alignment of clay minerals during deposition and compaction. Fissile mudrocks are often used in construction and as a source of hydrocarbons.

3. **Color Variability**: Mudrocks can exhibit a wide range of colors, including gray, black, brown, and red. The color is often influenced by the mineral composition, organic content, and the conditions under which the sediments were deposited. For example, dark-colored mudrocks may indicate a high organic content, while red mudrocks may suggest oxidation of iron minerals.

4. **Porosity and Permeability**: Mudrocks generally have low porosity and permeability due to their fine-grained nature. The small particle size and compact arrangement of grains limit the spaces available for fluid movement. However, certain types of mudrocks, such as shales, can serve as effective reservoirs for oil and gas, as well as barriers to fluid flow in subsurface environments.

5. **Organic Content**: Many mudrocks contain significant amounts of organic material, which can contribute to the formation of fossil fuels. The accumulation of organic matter in anoxic environments, such as swamps or deep marine settings, can lead to the generation of hydrocarbons over geological time.

Importance of Mudrocks

Mudrocks play a vital role in various geological and environmental processes:

- **Resource Reservoirs**: Mudrocks, particularly shales, are important sources of hydrocarbons, including oil and natural gas. The organic-rich layers can generate and trap hydrocarbons, making them significant in energy production.

- **Soil Formation**: The fine particles in mudrocks contribute to soil formation and fertility. As mudrocks weather, they release nutrients that can enhance soil quality, supporting vegetation and agriculture.

- **Sedimentary Environments**: Mudrocks provide insights into past environmental conditions and sedimentary processes. Their presence in the geological record can indicate low-energy depositional environments, such as lakes, deltas, and deep marine settings.

- **Construction Materials**: Certain types of mudrocks, such as shales, are used in construction and manufacturing. They can be processed into bricks, tiles, and other building materials due to their durability and availability.

CHAPTER FIVE

CHARACTERISTICS FOR NON-SILICICLASTIC ROCKS

Genesis: Sedimentary rocks are primarily formed through processes of sedimentation and diagenesis, which involve the accumulation, compaction, and cementation of mineral particles. The minerals present in these rocks are often derived from pre-existing rocks or are formed through chemical processes in the environment. Among the various types of sedimentary rocks, carbonate rocks are particularly significant due to their composition and formation processes.

Carbonate Rocks

Carbonate rocks are predominantly composed of carbonate minerals, which include:

- **Calcite ($CaCO_3$)**: The most common carbonate mineral, calcite is a key component of many sedimentary rocks, including limestone. It can form through both biological and inorganic processes.
- **Aragonite ($CaCO_3$)**: Aragonite is another form of calcium carbonate that is often found in marine environments. It is less stable than calcite and can transform into calcite over geological time.
- **Dolomite ($CaMg(CO_3)_2$)**: Dolomite is formed from the alteration of limestone through a process known as dolomitization, where magnesium ions replace some of the calcium ions in calcite.

In carbonate rocks, both the cement (the material that binds the clasts together) and the clasts themselves (which can include fossils and voids) can consist of these carbonate minerals. The presence of fossils in carbonate rocks often indicates a biological origin, as many carbonate rocks form in marine environments where organisms contribute to sedimentation.

Classification of Non-Siliciclastic Rocks

Non-siliciclastic sedimentary rocks can be categorized into three main groups: chemical, biochemical, and miscellaneous sedimentary rocks.

1. **Chemical Sedimentary Rocks**:

These rocks form when mineral constituents in solution become supersaturated and crystallize inorganically (Table 3.2).

. This process can occur in various environments, such as lakes, oceans, and hot springs. Examples of chemical sedimentary rocks include:

Oolitic Limestone: Formed from the accumulation of ooids, which are small, spherical grains composed of concentric layers of calcium carbonate. Oolitic limestone typically forms in warm, shallow marine environments.

Evaporites: These rocks form from the evaporation of water, leading to the precipitation of minerals. Common evaporite minerals include:

Rock Salt (Halite, NaCl): Formed from the evaporation of seawater or saline lakes.

Gypsum ($CaSO_4 \cdot 2H_2O$): Another evaporite mineral that forms in similar conditions as rock salt.

Barite ($BaSO_4$): Often forms in marine environments and can precipitate from seawater.

Sylvite (KCl): A potassium-bearing evaporite mineral that forms in saline environments.

2. **Biochemical Sedimentary Rocks**:

These rocks are formed from the accumulation of organic materials, such as shells, coral, and plant debris. The biological activity of organisms plays a significant role in the formation of these rocks. Examples include:

Limestone: Often formed from the accumulation of calcite shells from marine organisms, such as mollusks and corals.

Coal: Formed from the accumulation of plant material in swampy environments, where anaerobic conditions prevent complete decomposition.

3. **Miscellaneous Sedimentary Rocks**:

This category includes sedimentary rocks that do not fit neatly into the chemical or biochemical classifications. Examples may include certain types of conglomerates or breccias that contain a mix of different materials.

Conclusion

Sedimentary rocks, particularly carbonate rocks, play a vital role in understanding geological processes and the history of the Earth. The formation of these rocks through sedimentation and diagenesis involves complex interactions between biological, chemical, and physical processes. By studying the composition and classification of sedimentary rocks, geologists can gain insights into past environments, climate conditions, and the evolution of life on Earth. The ongoing processes of sedimentation and mineral crystallization continue to shape the Earth's surface, contributing to the dynamic nature of our planet's geology.

Table 3.2: Major Evaporites

Rock name	**Composition**	**Texture**
Crystalline Limestone	Calcite – $CaCO_3$	Coarse to fine crystalline
Fossiliferous Limestone	Calcite - $CaCO_3$	Visible fragments of shells
Chert	Quartz –SiO_2	Very fine crystalline
Gypsum	Gypsum – $CaSO_4 \cdot 2H_2O$	Fine to coarse crystalline
Chalk	Calcite - $CaCO_3$	Microscopic shells and clay
Rock salt	Halite -NaCl	Fine to coarse crystalline

Biochemical sedimentary rocks: Biochemical sedimentary rocks evolved when organisms use materials dissolved in air or water to build their tissue and become transformed into bedrocks when the organisms die. Examples include;

- Most types of limestones that were formed from the calcareous skeletons of organisms such as corals, mollusks, and foraminifera.
- Coal which formed from plants. The plants removed carbon from the atmosphere and combined it with other elements to build their tissues.
- Deposits of chert formed from the accumulation of siliceous skeletons of microscopic organisms such as radiolaria and diatoms.

These rocks are less resistant to chemical attack unlike siliciclastic rocks for which mineral constituents have strong silicate structures.

Miscellaneous sedimentary rocks

The miscellaneous category of sedimentary rocks encompasses those formed by processes that do not fit neatly into the traditional classifications of clastic, chemical, or biochemical sedimentary rocks. This category includes a variety of unique geological formations resulting from pyroclastic flows, impact events, volcanic activity, and other relatively uncommon processes. These rocks often exhibit distinctive characteristics that reflect their diverse origins and the complex interactions of geological forces.

Examples:

Pyroclastic Flows: These are fast-moving currents of hot gas and volcanic matter that flow down the slopes of a volcano during an explosive eruption. The materials ejected during these flows can include ash, pumice, and larger volcanic fragments. When these materials settle and lithify, they can form sedimentary deposits that reflect the violent nature of volcanic activity. Pyroclastic deposits can vary in texture and composition, depending on the size and type of volcanic material involved.

Impact Breccias: Formed from the shattering of rocks due to high-energy impacts, such as meteorite strikes, impact breccias consist of angular fragments that are cemented together. The intense pressure and heat generated during an impact can break apart surrounding rock, leading to the formation of these unique sedimentary rocks. Impact breccias provide valuable insights into the history of impacts on Earth and other planetary bodies, as they often contain shocked minerals and other indicators of high-energy events.

Volcanic Breccias: These rocks are composed of broken fragments of volcanic material, including both lava and pyroclastic components. Volcanic breccias can form during explosive eruptions or as a result of the collapse of volcanic structures. The angularity of the clasts in volcanic breccias indicates that they have not been transported far from their source, preserving their sharp edges.

Sedimentary Rocks of Mixed Percentages: This category includes rocks that may contain a mixture of clastic, chemical, and organic components, making them difficult to classify within the traditional sedimentary rock categories. These rocks can form in environments where multiple sedimentary processes occur simultaneously, leading to complex compositions that reflect the interplay of different geological forces.

Other Relatively Uncommon Processes: This may include rocks formed by unique geological events such as landslides, mass wasting, or the accumulation of materials in unusual environments. These processes can lead to the formation of sedimentary rocks that exhibit distinct characteristics and provide insights into specific geological conditions.

Conclusion

The classification of sedimentary rocks is fundamental to understanding their origins, characteristics, and the processes that shape them. By employing both compositional and genetic classification schemes, geologists can uncover valuable insights into the geological history of a region, the environmental conditions present at the time of deposition, and the potential resources these rocks may harbor. This comprehensive approach to classification not only enhances academic study but also has significant practical implications across various fields, including resource exploration, environmental geology, and construction.

Understanding the diverse categories of sedimentary rocks, including the miscellaneous group, enables geologists to appreciate the complexity of geological processes and the dynamic nature of the Earth's crust. Each category of sedimentary rock reflects specific depositional environments and geological histories, providing a narrative of the Earth's past. For instance, the presence of certain sedimentary rock types can indicate ancient marine environments, river systems, or arid conditions, allowing geologists to reconstruct past landscapes and climate conditions.

This knowledge is crucial for effective resource management, as it informs strategies for sustainable extraction and utilization of natural resources. Sedimentary rocks often serve as reservoirs for vital resources such as groundwater, hydrocarbons, and minerals. By understanding the characteristics and distribution of these rocks, geologists can develop more efficient methods for resource exploration and extraction, minimizing environmental impact and promoting sustainability.

Additionally, sedimentary rock classification is essential for environmental assessment. As human activities increasingly affect geological formations and ecosystems, understanding the composition and structure of sedimentary rocks allows for the evaluation of potential consequences. For instance, knowledge of sedimentary processes can guide land-use planning, construction practices, and the management of natural hazards like landslides and erosion.

Furthermore, advancements in geological sciences depend on a solid understanding of sedimentary rock classification. As researchers explore the intricate relationships between sedimentary processes and the Earth's history, this foundational knowledge will aid in the development of geological theories and practices. The integration of modern technologies, such as remote sensing and geospatial analysis, with traditional geological methods enhances our capacity to study sedimentary rocks and their environments.

Ultimately, a solid understanding of sedimentary rock classification enhances our knowledge of the Earth and helps address challenges related to geological resources. Insights from these studies are crucial for promoting sustainable practices amid global issues like climate change and resource depletion. Continued research and education in sedimentary geology will help us navigate the relationship between human activity and the natural world, fostering a sustainable future for generations to come.

CHAPTER SIX

MAJOR PHYSICAL PROPERTIES OF SEDIMENTARY ROCKS

Sedimentary rocks exhibit a range of physical properties that are crucial for understanding their formation, classification, and practical applications. These properties include texture, color, porosity, permeability, density, and hardness.

- **Texture** refers to the size, shape, and arrangement of the grains or particles within the rock, which can provide insights into the depositional environment and the processes that formed the rock.
- **Color** can indicate the mineral composition and the conditions under which the rock was formed, with variations often reflecting the presence of organic material, iron oxides, or other minerals.
- **Porosity** is the measure of void spaces within the rock, which affects its ability to hold fluids, while **permeability** refers to the rock's capacity to transmit fluids through its pore spaces.
- **Density** is an important property that can influence the rock's stability and behavior under stress, while **hardness** indicates the rock's resistance to abrasion and weathering.

Understanding these major physical properties is essential for geologists and engineers alike, as they play a significant role in resource exploration, environmental assessments, and construction practices.

(a) Color: The color of a sedimentary rock is often mostly determined by iron, an element of two major oxides i.e. iron (II) oxide and iron (III) oxide. Iron (II)oxide only forms under anoxic circumstances and gives the rock a grey or greenish colour. Iron (III) oxide is often in the form of the mineral hematite and gives a rock reddish to brownish colour. In arid continental climates rocks are in direct contact with the atmosphere, and oxidation is fairly active giving the rock a red or orange colour as exhibited by 'red beds'. However, the red coloration is not restricted to such environments alone.

The presence of organic material can colour a rock black or grey. Organic material is formed in nature from mostly dead plants and other macro-organisms. Normally, such material eventually decays by oxidation. The material can also decay by bacterial activity. At the bottom of deep seas and lakes, there is little water current in such environments, no oxygen from surface water is brought down to facilitate oxidation; hence deposited sediments remained darkish in color such as the usually fine dark grey color for shale deposits.

Typical banded ironstones, such as the one illustrated in figure 3.19, consist of alternating layers of iron (III) oxide (red) and iron (II) oxide (grey). The banded ironstones were formed mostly during Precambrian times (about 3700 MA) when the atmosphere had a low concentration of oxygen compared to other gasses.

Fig.3.19: Banded Iron Formation: Typical rock sample-Inset(A) and Field Exposure(B) Karijini National Park, Western Australia

(b) Sedimentary Textures: Sedimentary texture refers to size, form and orientation of clasts or minerals that constitute the rock. The texture is a small-scale property of the rock, but contributes to large-scale properties, such as the density, porosity or permeability. Clastic rocks have a 'clastic texture', which means they consist of clasts predominantly. Matrix or cement occurs between the clasts. Cement which consists of crystals of one or more mineral types and being post depositional mineral growth binds the clasts and other smaller fragments. The size and form of clasts constituting sediments can be used to determine the velocity and direction of current in the sedimentary environment where the rock was formed. Thus, fine, calcareous mud only settles in calmed water, while gravel and larger clasts are only deposited by rapidly moving water.

(c) Grain size and Sorting: Usually, the grain size of a rock is categorized based on the Wentworth scale although alternative scales can also be used. The grain size can be expressed as a diameteror as volume, and is always an average value since clasts differin sizes. Statistical distribution of grain sizes is normally different for different rock types and is considered as a characteristic for sorting of rock fragments. Sorting as applied in sedimentary rocks refers to the degree of uniformity of grain size. The degree of sorting depends upon the distance that is covered during transportation. A 'very poorly sorted' sedimentary rock indicates that the fragments are mixed in grain sizes and covered a very short distance while 'well sorted' indicates that the fragments have similar grain sizes and covered a longer distance. The degree of sorting increases with increase in the degree of rounding and degree of sphericity (Fig.3.20). The figure shows five degrees of sorting in 'A' and poorly sorted Cobbles and Pebbles in layer C and well sorted fine grained sand of layer F in 'B'. Rounded particles when transported roll faster than irregularly shaped fragments thus having a better chance to be well sorted out based on particle grain size.

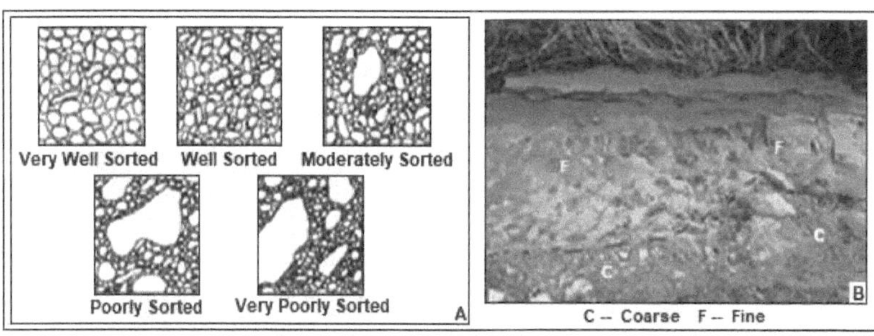

Fig.3.20: Sorting Illustrations(A) and Field Exposure for Sorting

(d) Form: Form refers to the nature of the surface and shape for clasts. It can reflect the formation process of the rock. For instance, Coquina is a rock that consists of cladsts of broken shells. It formed under energetic water conditions. The form for a clast can be described by using any of the following four parameters:

- *Surface texture: The* amount of small-scale relief of the surface of a grain that is too small to influence the general shape.
- '*Grain form*': The term describes the three dimensional shape of the grain.Chemical sedimentary rocks consist entirely of crystals. Only average size of crystals and fabric are considered in textural descriptions.
- '*Sphericity*': The degree to which a grain approaches a sphere. A long

distance is need during transportation to achieve the spherucity.

- Rounding, also known as angularity, is the degree to which the individual sedimentary particles are rounded. The longer the distance the grain gets transported, the more the corners of an individual grain becomes rounded (Fig.3.21).

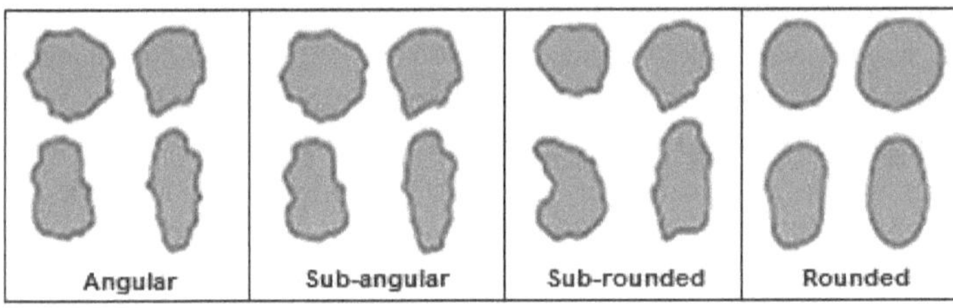

Fig.3.21: Rounding

(e) Fossils: Fossils as preserved hard part of once living organism are prevalent in calcareous sedimentary rocks. Unlike most igneous and metamorphic rocks, sedimentary rocks form at favorable conditions that preserve macro- and micro-fossils. Dead organisms in nature are usually quickly removed by scavengers, bacteria and erosion. However, sedimentation processes can create exceptional circumstances that lead to fossilization. Chances of fossilization is higher when sedimentation rate is high for fast burial of fossil candidates thereby eliminating bacterial activities. Larger, well-preserved fossils are relatively rare although most commonly preserved are the harderparts of organisms such as bones, shells, woody tissue of plants. However, a soft tissue has a much smaller chance of being preserved and fossilized. Imprints of organisms made while still alive are called trace fossils. Examples are burrows, footprints, etc. Fossils can be both direct remains and imprints of organisms or their skeletons. Being part of other sedimentary material, fossil undergo the same diagenetic processes. For instance, a shell consisting of calcite can get dissolved but be precipitated later so as to fill the void created.

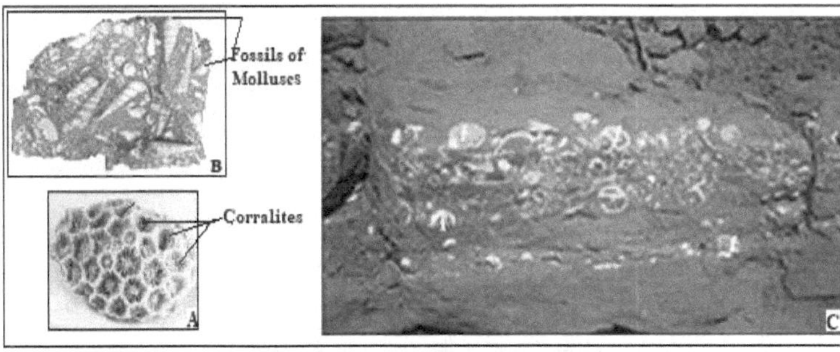

Fig.3.22: Typical Fossils in (A) and in (B), and Fossilifrous Sedimentary Rock Exposure(C) of Ano Nuevo State Reserve, California

Typical fossils, as illustrated in Figure 3.22, provide valuable insights into the history of life on Earth. Fossils are the preserved remains or traces of ancient organisms, and they can form through various processes. One significant method of fossilization is known as **permineralization**, which occurs when minerals precipitate from groundwater and fill the empty spaces within the remains of an organism. This process not only preserves the shape of the original organism but also alters its chemical composition.

During permineralization, the original organic material is gradually replaced by minerals, resulting in a fossil that retains the external morphology of the organism while undergoing significant changes in its internal structure. This transformation allows for the preservation of fine details, such as cellular structures, which can be crucial for paleontological studies.

The most common minerals involved in the cementation process during permineralization include:

- **Carbonates**: The most prevalent carbonate mineral is **calcite** ($CaCO3$), which can precipitate from calcium-rich waters. Calcite is often found in fossils of marine organisms, such as mollusks and corals, where it can replace the original aragonite or other calcium carbonate forms.

- **Amorphous Silica**: This includes minerals such as **chalcedony**, **flint**, and **chert**. These silica forms can replace organic material in plants and some invertebrates, leading to the preservation of intricate details. Chalcedony, for instance, is a microcrystalline form of silica that can fill in the spaces of plant tissues, resulting in beautifully preserved fossilized wood known as petrified wood.

- **Pyrite**: Often referred to as "fool's gold," pyrite (FeS_2) can also play a role in permineralization. It forms in anoxic environments where sulfur is abundant, and it can replace organic material, leading to the preservation of fossils with a metallic luster. Pyritized fossils can provide important information about the conditions under which they formed.

The process of permineralization is essential for understanding the fossil record, as it allows for the preservation of organisms that lived millions of years ago. By studying these fossils, paleontologists can gain insights into ancient ecosystems, evolutionary processes, and the environmental conditions that existed during different geological periods. Fossils formed through permineralization not only enrich our understanding of biological history but also contribute to the broader field of geology by providing context for sedimentary rock formation and the conditions of past environments.

CHAPTER SEVEN

SEDIMENTARY STRUCTURES

(a) **Primary sedimentary structures:** Structures in sedimentary rocks can be divided into 'primary' structures (formed during deposition) and 'secondary' structures (formed after deposition). Unlike textures, structures are always large-scale features that can easily be studied in the field. Sedimentary structures can unveil something about the sedimentary environment or can serve to reveal parts of deposits that originally faced up wheretectonics will have tilted or overturned the sedimentary layers. Usually, sedimentary rocks are laid down in layers called beds or strata. A bed is defined as a layer of rock that has a uniform lithology and texture. A single bed can be a couple of centimeters to several meters thick. Beds form by the deposition of layers of sediment on top of each other. The sequence of beds that characterizes sedimentary rocks is called bedding. Finer, less pronounced layers are called laminae and the stratification of laminae is known as lamination.

Laminae are usually less than a few centimeters thick. Although beddings and laminations are often originally horizontal in nature, this is not always the case since after initial layering is over, the layers are subjected to deformation etc.

Graded bedding: Graded bedding as illustrated for unconsolidated deposits in figure 2.23(A) is a structure where beds with a smaller grain size occur on top of beds with larger grains. This structure forms when fast flowing water, as a transporting agent stops to flow. The stoppage is followed first by settling of larger and heavier clasts and then followed by smaller clasts. The conditions of deposition can also reoccur leading to a similar type of layering (repetitive).

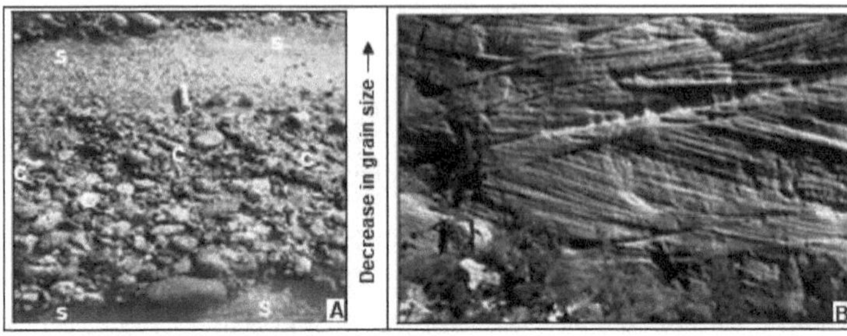

Fig.3.23: Typical Graded bedding (A) and Jurassic Eolian Cross-Bedded Sandstone(B), Zion National Park, Utah

Cross bedding: In some environments, beds are deposited usually at small angles to the horizontal. Sometimes multiple sets of layers with different orientations exist in the same rock thereby forming, a structure known as cross-bedding. The cross-bedding as shown in figure 3.23(B) forms when an old bed is eroded and becomes tectonically disturbed so as to attain a different dip. Thereafter, material for a younger bed are emplaced onto the erosional surface. The opposite of cross-bedding is known as parallel lamination, where all sedimentary layering is parallel. Differences for laminations are generally caused by cyclic changes in the sediment supply. The changes may involve seasonal changes in rainfall, temperature, biochemical activity etc. Some rocks have no lamination at all and their structural character is called massive bedding.

Bedforms: The term 'Bedform' refers to the nature of the top surface for a particular bed. It can also be indicative for dynamics that occurred for a particular sedimentary formation. Examples of bed forms include ripple marks for stream sediments and sand dunes, Flute casts and mudcraks.

Ripple marks: Ripple marks develop on sediments that are deposited by flowing water or by wind activity in arid or semi-arid regions. Ocean currents also deposit fine sand on beaches to build up sand dunes. The sand dunes also develop ripples. At times of low tide, the water currents subside and water waves are able to creat the ripple marks. There are two type of ripple marks i.e. asymmetric wave ripples and symmetric current ripples. Deposition that involves the current in one direction, such as for rivers, produces asymmetric ripples (Fig.3.24 (A)). The longer flank of such ripples is oriented in the direction that is opposite to the direction of the current. Wave ripples occur in environments where currents occur in all directions, such as tidal flats and can be well preserved (Fig.3.24 (B)).

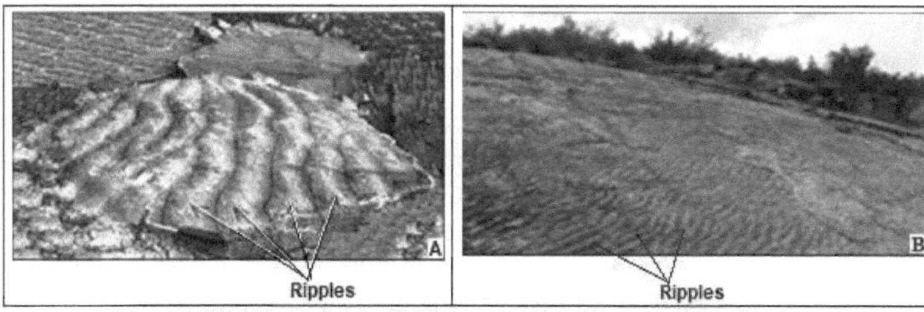

Fig.3.24: Ripple marks on Habberg Sandstones, Bavaria, Germany and Rippled Cretaceous Dakota Formation, Colorado

Flute casts: Features such as flute casts, are groves that were dug into a sedimentary layer and then preserved. They are often elongated structures and can be used to establish the direction of the flow during deposition. The grooves dry up and beome filled by fine sediments of a different material. Furthermore, the overlying rock pressure has to be low to safe the tracks from being deformed.

Mud cracks: Mudcracks are caused by dehydration of sediment that occasionally gets exposed to the atmosphere especially in tidal flats or point bars along rivers. In such environments, fine sediments contain a high concentration of clay minerals other than silt size fragments. As is usual behavior, clay minerals become expansive when in contact with water. Conversely the clays shrink when drying. The effect of drying is manifested in the development of weak zones. The weak zones developed form a pattern that is polygonal as shown in figure 3.25. Development of mud cracks is also experienced in expansive soils. The expansive soils are always a challenge in construction works. However, engineering solutions have been determined to address the challenges involved.

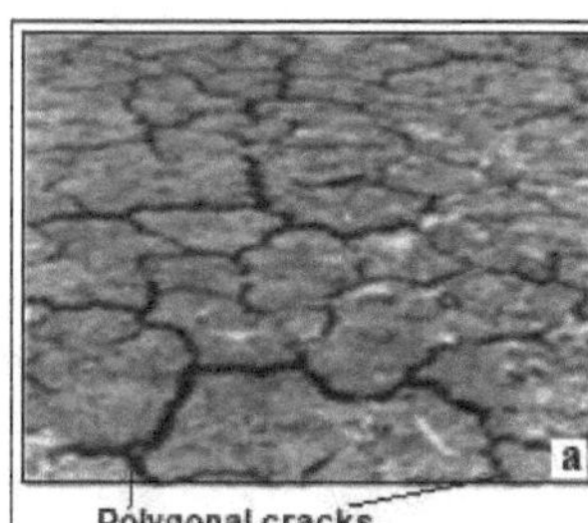

Fig. 3.25: Typical Tidal flats Mudcracks

(b)Secondary sedimentary structures: Secondary sedimentary structures are features that develop in sedimentary rocks after the initial deposition of sediments. Unlike primary structures, which form during the deposition process, secondary structures arise from various chemical, physical, and biological processes that occur within the sediments over time. These structures provide valuable insights into the environmental conditions and biological activities that existed in the past.

One notable example of secondary sedimentary structures is **trace fossils** (or ichnofossils), which are preserved tracks or traces left by organisms as they moved across or interacted with the sediment. These can include:

- **Preserved Paw Prints**: Fossilized footprints of dinosaurs and other animals provide critical information about their size, gait, and behavior, as well as insights into the ecosystems they inhabited.
- **Burrows**: The burrows of mollusks, arthropods, and other organisms can also be preserved as secondary structures. These burrows indicate the presence of life and can reveal details about the sedimentary environment, such as its stability and the types of organisms that lived there.

These secondary structures are valuable indicators of the biological and ecological environments that existed in the past. By studying them, paleontologists and geologists can reconstruct ancient ecosystems, understand organism behavior, and infer the conditions under which the sediments were deposited.

Fig.3.26: Chert Concretions (c)in Chalk of Middle Lefkara Formation, Cyprus

In addition to trace fossils, secondary sedimentary structures can also result from the **precipitation of certain chemical species** within the sediments. This process can lead to:

- **Coloring and Staining**: The precipitation of minerals can alter the color of the rock, providing clues about the chemical environment during and after deposition. For example, iron oxides can impart reddish or yellowish hues to sedimentary rocks.
- **Formation of Concretions**: Concretions are roughly concentric bodies that have a different composition from the surrounding host rock (Fig. 3.26). They form when minerals precipitate around a nucleus, often leading to the development of distinct shapes and sizes. Concretions can vary in composition, often consisting of minerals such as calcite, quartz, or iron oxide, and can provide insights into the diagenetic processes that occurred after sediment deposition.

Overall, secondary sedimentary structures are essential for deciphering the geological history of an area. These structures provide valuable insights into past biological activity, revealing how organisms interacted with their environment and the conditions that prevailed during different geological periods. For instance, trace fossils such as footprints and burrows not only indicate the presence of life but also offer clues about the behavior, movement patterns, and ecological roles of ancient organisms. By analyzing these structures, paleontologists can reconstruct the dynamics of prehistoric ecosystems, including predator-prey relationships and the types of habitats that existed.

In addition to biological insights, secondary sedimentary structures enhance our understanding of sedimentary processes. They illustrate how sediments can be altered after deposition through various mechanisms, including chemical precipitation, compaction, and bioturbation. For example, the formation of concretions—mineral masses that develop around a nucleus—can indicate localized changes in chemical conditions within the sediment, such as variations in pH or the availability of certain ions. These changes can inform geologists about the diagenetic processes that occurred after the initial deposition, shedding light on the conditions that influenced the transformation of sediments into rock.

Moreover, secondary sedimentary structures can also provide information about past environmental conditions, such as water depth, energy levels, and sedimentation rates. For instance, the presence of certain types of burrows may suggest a stable substrate conducive to life, while the characteristics of concretions can indicate periods of increased mineral precipitation, possibly linked to changes in water chemistry or biological activity. By studying these structures, geologists can infer the depositional environments of sedimentary rocks, whether they formed in marine, fluvial, or terrestrial settings.

The significance of secondary sedimentary structures extends beyond academic interest; they have practical implications in fields such as resource exploration and environmental management. For example, understanding the distribution and characteristics of these structures can aid in locating fossil fuel deposits, groundwater reservoirs, and mineral resources. Additionally, insights gained from studying these structures can inform land-use planning and environmental assessments, helping to mitigate the impacts of human activities on geological formations and ecosystems.

In summary, secondary sedimentary structures are crucial for understanding the complex interplay between biological activity, sedimentary processes, and environmental conditions throughout geological history. They not only enrich our knowledge of past life and ecosystems but also contribute to our understanding of the dynamic evolution of Earth's surface over time. By integrating the study of these structures into geological research, we can gain a more comprehensive view of the Earth's history and its ongoing changes, ultimately enhancing our ability to manage and protect our planet's resources and environments.

CHAPTER EIGHT

COMPARATIVE ILLUSTRATIONS

8.1 Siliciclastic Rocks

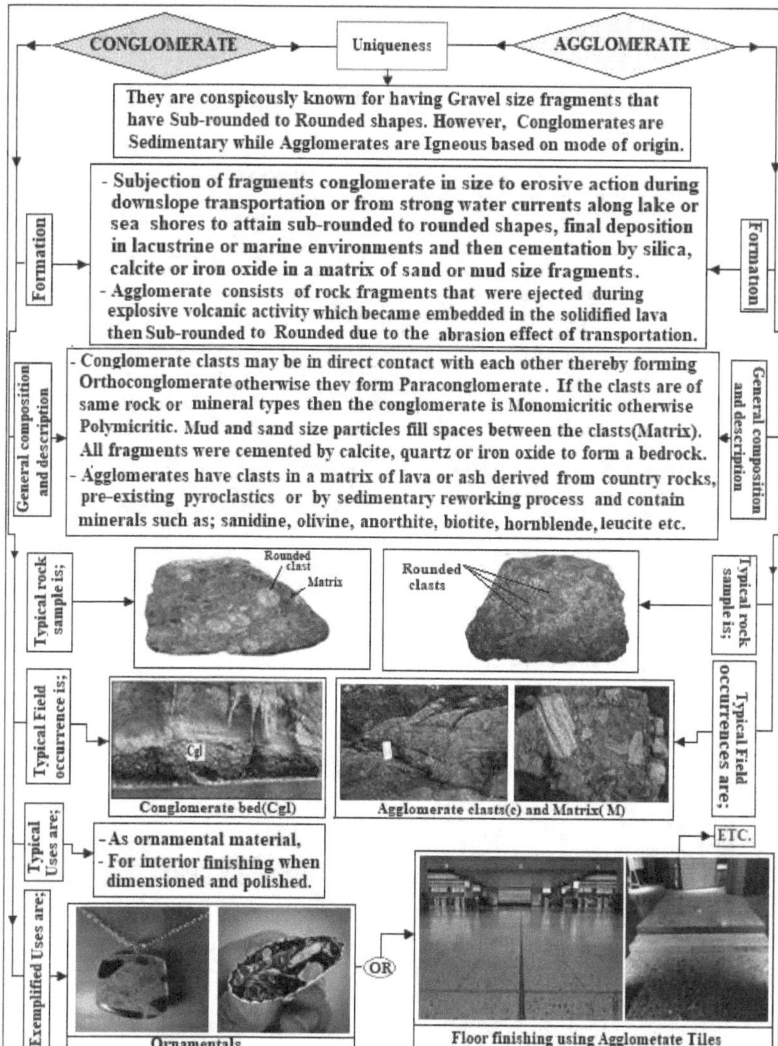

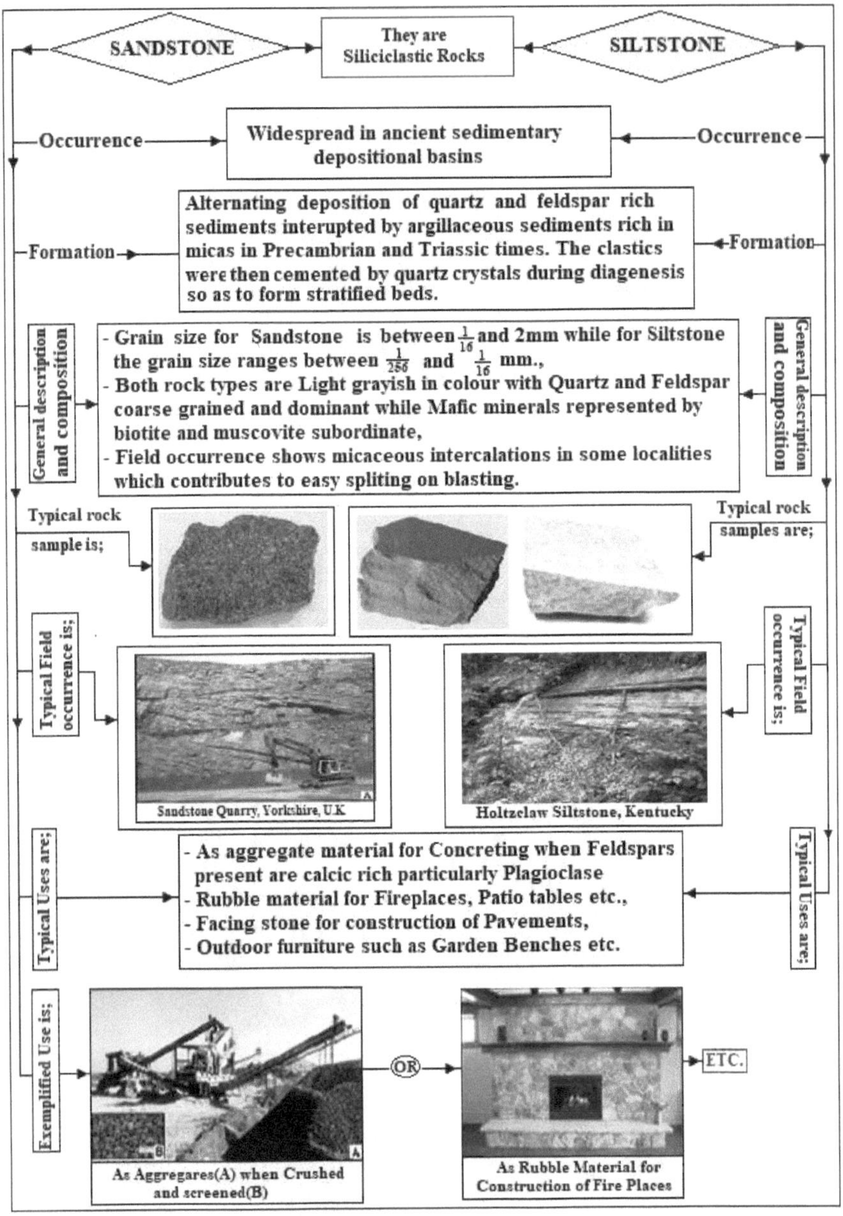
SANDSTONE
They are Siliciclastic Rocks
SILTSTONE
Occurrence
Widespread in ancient sedimentary depositional basins
Occurrence
Formation
Alternating deposition of quartz and feldspar rich sediments interupted by argillaceous sediments rich in micas in Precambrian and Triassic times. The clastics were then cemented by quartz crystals during diagenesis so as to form stratified beds.
Formation
General description and composition
- Grain size for Sandstone is between $\frac{1}{16}$ and 2mm while for Siltstone the grain size ranges between $\frac{1}{256}$ and $\frac{1}{16}$ mm.,
- Both rock types are Light grayish in colour with Quartz and Feldspar coarse grained and dominant while Mafic minerals represented by biotite and muscovite subordinate,
- Field occurrence shows micaceous intercalations in some localities which contributes to easy spliting on blasting.
General description and composition
Typical rock sample is;
Typical rock samples are;
Typical Field occurrence is;
Sandstone Quarry, Yorkshire, U.K
Holtzclaw Siltstone, Kentucky
Typical Field occurrence is;
Typical Uses are;
- As aggregate material for Concreting when Feldspars present are calcic rich particularly Plagioclase
- Rubble material for Fireplaces, Patio tables etc.,
- Facing stone for construction of Pavements,
- Outdoor furniture such as Garden Benches etc.
Typical Uses are;
Exemplified Use is;
As Aggregares(A) when Crushed and screened(B)
OR
As Rubble Material for Construction of Fire Places
ETC.

CLAYSTONE → **Uniqueness** ← **SHALE**

Fine grained argillaceous rocks that are expanvive on absoption of water but when dried up, they have high shrinkage limits accompanied by high dry strength.

Formation

- Claystone sediments were derived mainly from granites and gneisses by weathering to form kaolinitic clays. The sediments were, in the geologic past deposited mostly in deltaic sections of rivers then transformed to Claystones.
- Sediments for shales being similar to clays but incoporated with organic matter were deposited, in alternating sequence, with silt and mud layers, in basins of shallow seas, river flood plains and playas then transformed to compact layers.

General Description and Composition

- Claystone is a dark gray to pinkish coloured sedimentary rock that is fine grained having particles grain size less than $\frac{1}{256}$mm in diam. Typical minerals contained are Kaolinite group which becomed Highly Plastic when in contact with water. Other minerals are quartz, feldspars, pyrite, micas, illite and oxides.
- A shale consists of clay size particles in greater proportion to silt size particles. Also, organic particles are in high proportion and contribute to the dark gray to black colour other than dark coloured minerals such as sulphides that could be present. Major minerals are quartz and feldspars while accessory minerals are iron oxide, sulphides, some carbonates and heavy minerals. The rock readily splits laterally into thin pieces on dehydration i.e. exhibits fissility property.

Typical rock sample is; (Claystone)

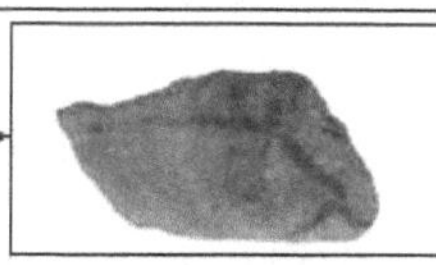

Typical rock samples are; (Shale)

Typical Field occurrence is; (Claystone)

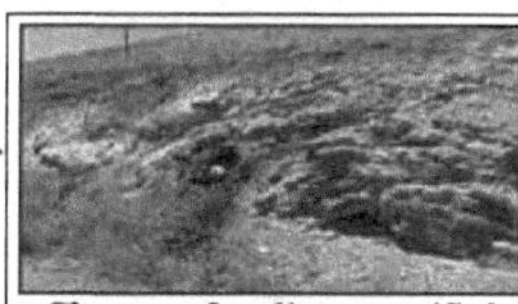

Claystone- Locality unspecified

Typical Field occurrence is; (Shale)

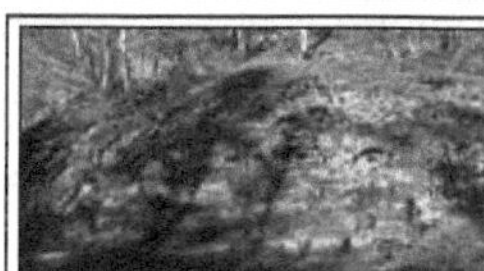

Bituminous Shale, Andrarum

Typical Uses are; (Claystone)

- Manufacture of Bricks,
- Cement manufacture,
- A sintering agent in Iron Processing
- Manufacture of mortar etc.
- Material for Pottery works.

Typical Uses are; (Shale)

- Manufacture of Bricks and Tiles,
- A material for Pottery Works,
- Cement maufacture when mixed with limestone and fired etc.

Exemplified Uses are;

As Bricks for Roofing(R), Walling (W) and Fencing (F)

and

As Material for Pottery

ETC.

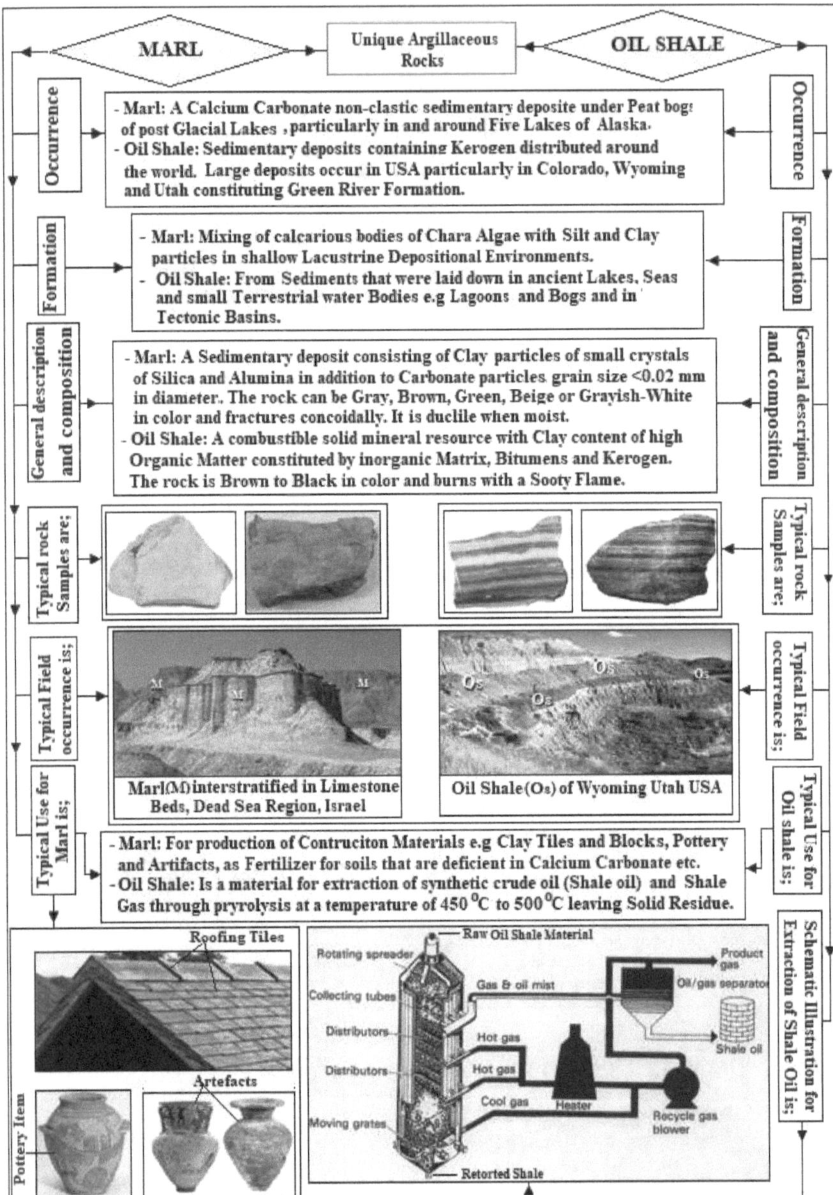
MARL
Unique Argillaceous Rocks
OIL SHALE
Occurrence
- Marl: A Calcium Carbonate non-clastic sedimentary deposite under Peat bogs of post Glacial Lakes ,particularly in and around Five Lakes of Alaska.
- Oil Shale: Sedimentary deposits containing Kerogen distributed around the world. Large deposits occur in USA particularly in Colorado, Wyoming and Utah constituting Green River Formation.
Occurrence
Formation
- Marl: Mixing of calcarious bodies of Chara Algae with Silt and Clay particles in shallow Lacustrine Depositional Environments.
- Oil Shale: From Sediments that were laid down in ancient Lakes, Seas and small Terrestrial water Bodies e.g Lagoons and Bogs and in Tectonic Basins.
Formation
General description and composition
- Marl: A Sedimentary deposit consisting of Clay particles of small crystals of Silica and Alumina in addition to Carbonate particles grain size <0.02 mm in diameter. The rock can be Gray, Brown, Green, Beige or Grayish-White in color and fractures concoidally. It is duclile when moist.
- Oil Shale: A combustible solid mineral resource with Clay content of high Organic Matter constituted by inorganic Matrix, Bitumens and Kerogen. The rock is Brown to Black in color and burns with a Sooty Flame.
General description and composition
Typical rock Samples are;
Typical rock Samples are;
Typical Field occurrence is;
M
M
M
Os
Os
Os
Os
Marl(M) interstratified in Limestone Beds, Dead Sea Region, Israel
Oil Shale(Os) of Wyoming Utah USA
Typical Field occurrence is;
Typical Use for Marl is;
- Marl: For production of Contruciton Materials e.g Clay Tiles and Blocks, Pottery and Artifacts, as Fertilizer for soils that are deficient in Calcium Carbonate etc.
- Oil Shale: Is a material for extraction of synthetic crude oil (Shale oil) and Shale Gas through pryrolysis at a temperature of 450 °C to 500 °C leaving Solid Residue.
Typical Use for Oil shale is;
Roofing Tiles
Artefacts
Pottery Item
Raw Oil Shale Material
Rotating spreader
Collecting tubes
Distributors
Distributors
Moving grates
Gas & oil mist
Product gas
Oil/gas separato
Shale oil
Hot gas
Hot gas
Cool gas
Heater
Recycle gas blower
Retorted Shale
Schematic Illustration for Extraction of Shale Oil is;

LATERITE → **Uniqueness** ← **CALCRETE**

They are natural deposits that are suitable and popular for soil compaction in Road Work Projects.

Formation

- Residual sedimentary deposits that form in tropical regions under hot and wet climatic conditions through chemical weathering that involves leaching of silicates from parent rocks accompanied by precipitation of insoluble oxides of Fe, Al and Ca, main source rocks being Basaltic and Rhyolitic lavas, and Gneisses
- Calcrete was formed as a result of fluctuation of climatic conditions in Arid and Semi-arid regions. In those regions, rainwater saturated with carbon dioxide from the atmosphere dissolved calcite minerals from calcarious material and got precipitated during drying conditions in the intestices of clastic fragments thereby acting as cement to form the hardpan layers.

General Description and Composition

- Laterite is dark brown to redish brown in colour rich in Iron and Aluminum. Typical minerals present are Quartz, Zircon, Oxides of Titanium, Iron, Aluminum and manganese.
- Calcrete also known as Hardpan is a calcarious rich duricrust material. It occurs as a hardened layer consisting of surficial sand and gravel material that was cemented by calcium carbonate.

Typical rock sample is;

Laterite material — Laterite quarry

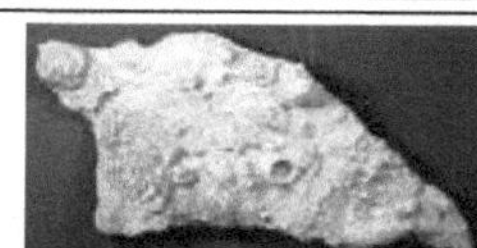

Typical Field occurrence is; / **Typical Field occurrences are;**

Lateritic (L) Landscape, Luanda Angola

Calcrete on San Miguel Island, California

Major Use is: Material for Field Compaction in Pavement Works such as;

Laterite Soil (LS) for Road Rehabilitation, Burera, Rwanda

-and-

Road construction at finishing stage, Niger

→ ETC.

BANDED IRONSTONE → **Banded Ironstone is a Siliceous Rock while Bentonit Clay is an Argillaceous Rock** ← **BENTONITE CLAY**

Occurrence

- Banded Ironstone: In Proteozoic Rocks all over the world oldest being Isua Greenstone Belt of Canada about 3700 to 3800 Ma years in Age.
- Bentonite Clay: In localities of Ancient Marine environments largest accumulation being in Fort Benton in Wyoming USA.

Formation

- Banded Ironstone: Reaction of iron with oxygen generated by Photosynthetic organisms which then dissolved in ocean water and accumulated alternately with Chert layers on Ocean Floors.
- Bentonite Clay: Accumulation of Volcanic Ash under Marine environment which then transformed into Clay deposits that were rich in Smectite minerals.

General description and composition

- Banded Ironstone:Consisits of Iron rich bands often Hematite Dark-Gray in color alternating with light-Redish Jaspilitic Chert Bands with some Goethite. The iron rich Bands contain > 60% by wt of Iron.
- Bentonite Clay: Consists of mainly Smectite Clay Minerals of Iron and Magnesium and either Sodium or Calcium. The Clay has high Swelling capacity when in contact with water. It also has high Viscosity and Thixotropy.

Typical rock Samples are;

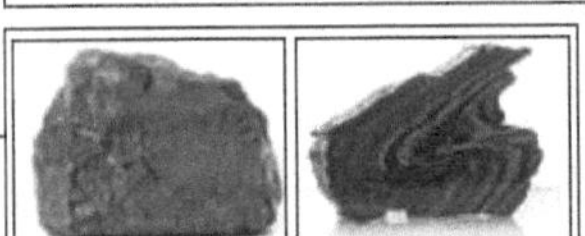

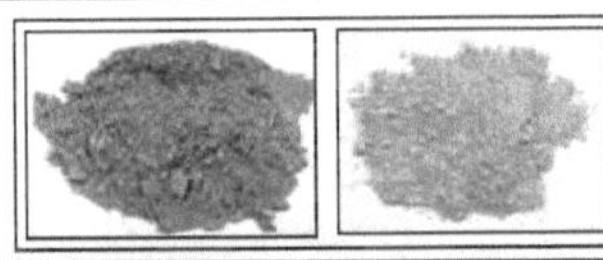

Typical Field Occurrence is;

Banded Ironstone at Karijini National Park, Western Australia.

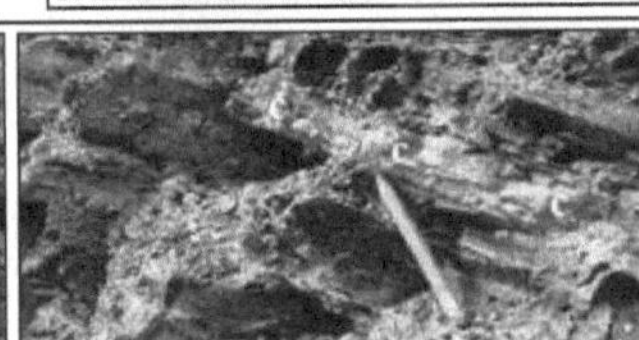

Slope exposure of Bentonite Clay(C) Untersberg, Germany.

Typical Uses for Banded Ironstone are; / Typical Uses for Bentonite Clay are

- Banded Ironstone: Raw material for extraction of Iron Metal and as Ornamental.
- Bentonite Clay: Flush Fluid in drilling for lubrication and lowering of temperature, A Bonding Agent in Foundry Moulding, A Medicinal material for Removal of Toxins from the body, Treatment of Oily Skin and Acne etc.

Examples for Uses of Bentonite Clay are;

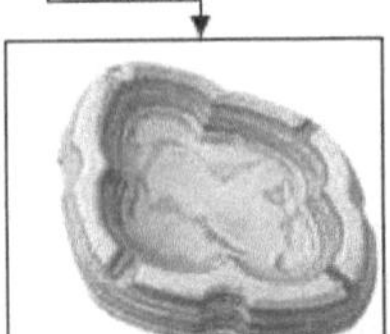

Ornamental Carving

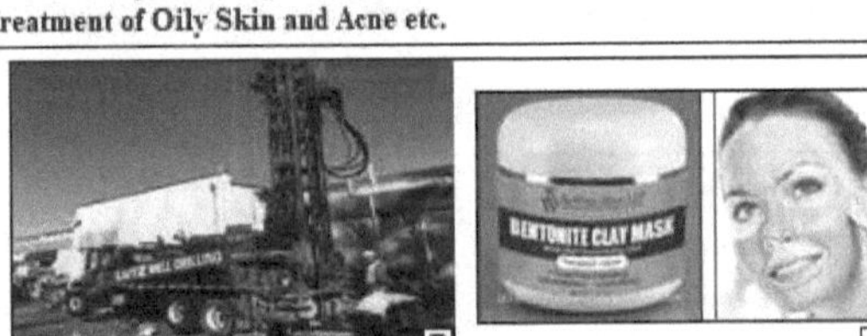

Skin Medication in (A) and Flush Fluid for Drilling in (B)

8.2 Non-Siliciclastic Rocks

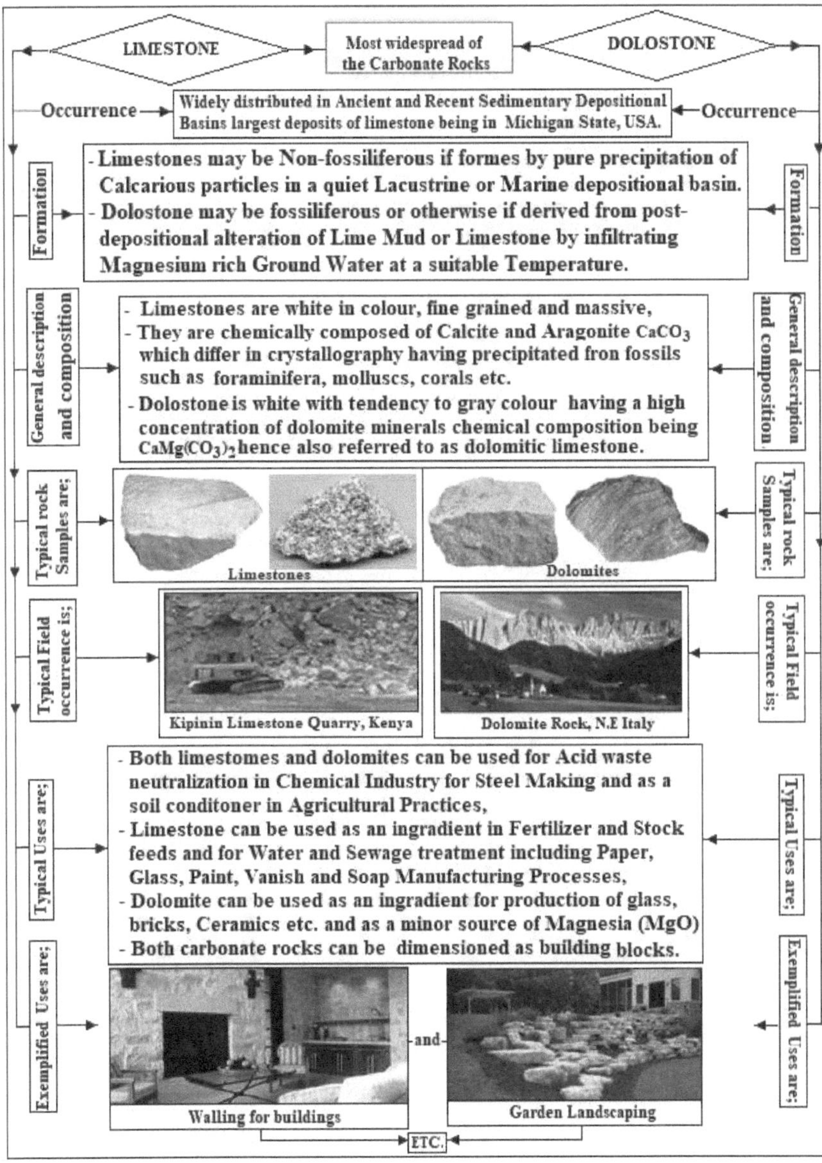

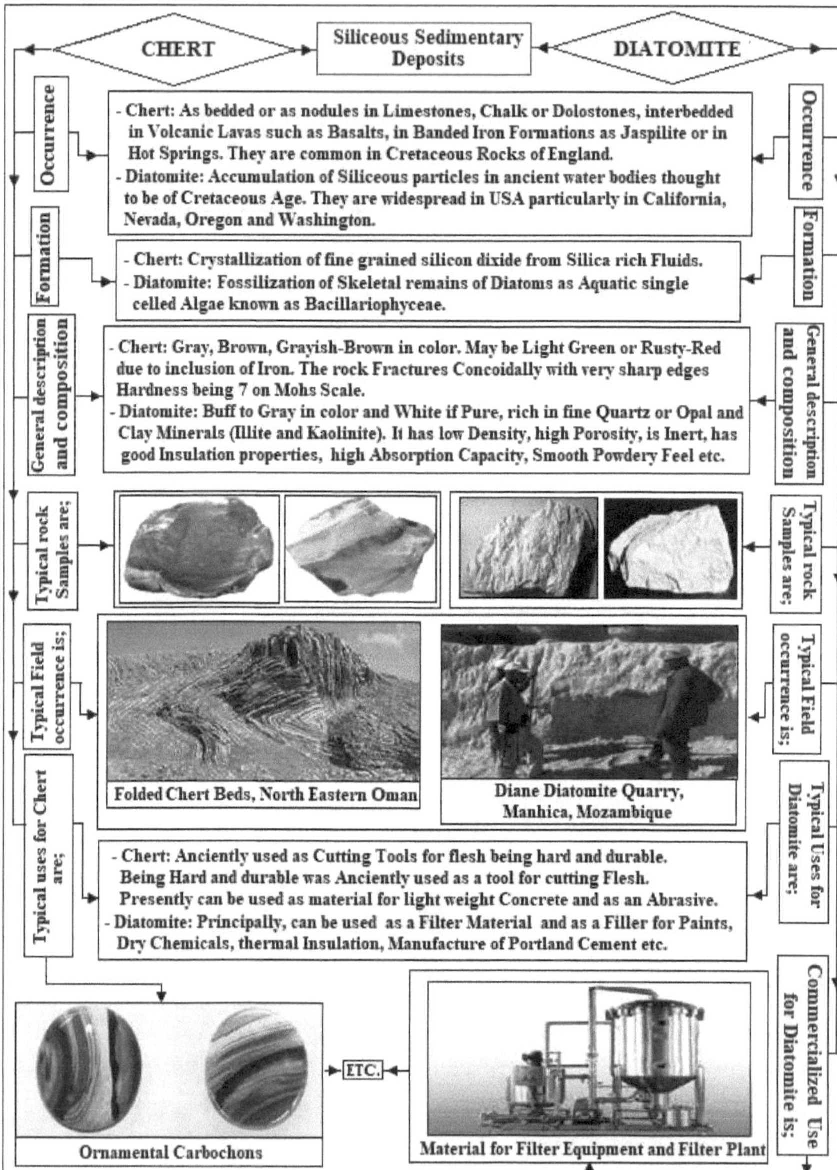
CHERT
Siliceous Sedimentary Deposits
DIATOMITE
Occurrence
- Chert: As bedded or as nodules in Limestones, Chalk or Dolostones, interbedded in Volcanic Lavas such as Basalts, in Banded Iron Formations as Jaspilite or in Hot Springs. They are common in Cretaceous Rocks of England.
- Diatomite: Accumulation of Siliceous particles in ancient water bodies thought to be of Cretaceous Age. They are widespread in USA particularly in California, Nevada, Oregon and Washington.
Formation
- Chert: Crystallization of fine grained silicon dixide from Silica rich Fluids.
- Diatomite: Fossilization of Skeletal remains of Diatoms as Aquatic single celled Algae known as Bacillariophyceae.
General description and composition
- Chert: Gray, Brown, Grayish-Brown in color. May be Light Green or Rusty-Red due to inclusion of Iron. The rock Fractures Concoidally with very sharp edges Hardness being 7 on Mohs Scale.
- Diatomite: Buff to Gray in color and White if Pure, rich in fine Quartz or Opal and Clay Minerals (Illite and Kaolinite). It has low Density, high Porosity, is Inert, has good Insulation properties, high Absorption Capacity, Smooth Powdery Feel etc.
Typical rock Samples are;
Typical Field occurrence is;
Folded Chert Beds, North Eastern Oman
Diane Diatomite Quarry, Manhica, Mozambique
Typical uses for Chert are;
Typical Uses for Diatomite are;
- Chert: Anciently used as Cutting Tools for flesh being hard and durable.
Being Hard and durable was Anciently used as a tool for cutting Flesh.
Presently can be used as material for light weight Concrete and as an Abrasive.
- Diatomite: Principally, can be used as a Filter Material and as a Filler for Paints, Dry Chemicals, thermal Insulation, Manufacture of Portland Cement etc.
Commercialized Use for Diatomite is;
Ornamental Carbochons
ETC.
Material for Filter Equipment and Filter Plant

LIGNITE — **Varieties of Coal Series** — **ANTHRACITE**

Occurrence

- Lignite: Widely distributed, highest reserves in Russia. Germany leading producer as at 2018 (About 32% of word's production). Other significant producers are China, Tamil Nadu, Poland , USA, South Africa.
- Anthracite: Widely distributed with largest reserves inPennyslvania State USA. Other considerable deposits are in South Africa, Australia, Ukrain, China etc.

Formation

By Partial decomposition of Plant Material that was subjected to increased Temperature over millions of years accompanied by dehydration.

General description and composition

- Lignite: Generally Yellow to Dark-Brown and rarely Black in color. Carbon content of 22-35%, Moisture content of 45-55%. Calorific value of 8000-17000KJ/Kg. High Sulfur content in form of Ferrous Sulphide.
- Anthracite: Black to Steel-Gray in color. Almost metallic Luster, High Density, Hard and Brittle. Low Moisture content of 5-15%but high Carbon content (80-95%) Has a higher Calorific value than Lignite i.e about 35000 MJ/Kg.

Typical rock Samples are;

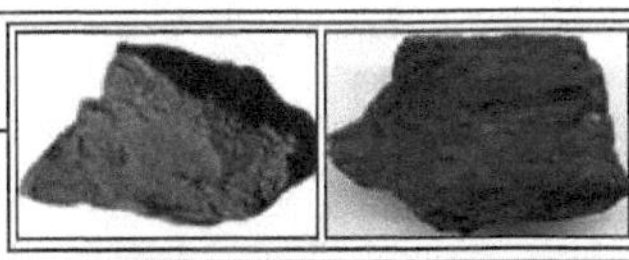

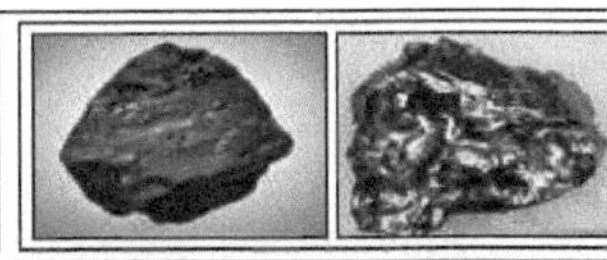

Typical Field occurrence is;

Gersweiler Surface Lignite Mine Germany

Anthracite Coal Mine, Sangamon County, Illinois State, USA

Major Uses for Lignite Coal are; / Major Uses for Anthracite Coal are;

Furnace heating in Metallurgical processes, Heavy chemistry and Railroad Transpotation. Production of Steam for Electrical Power Production.

Schematic Illustration for Lignite or Anthracite Powered Generation of Electricity is;

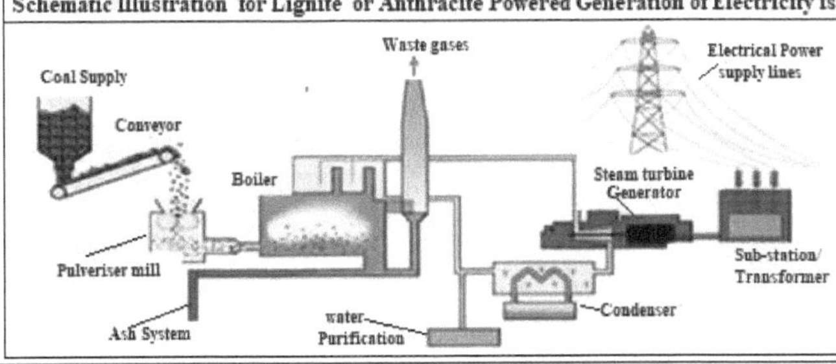

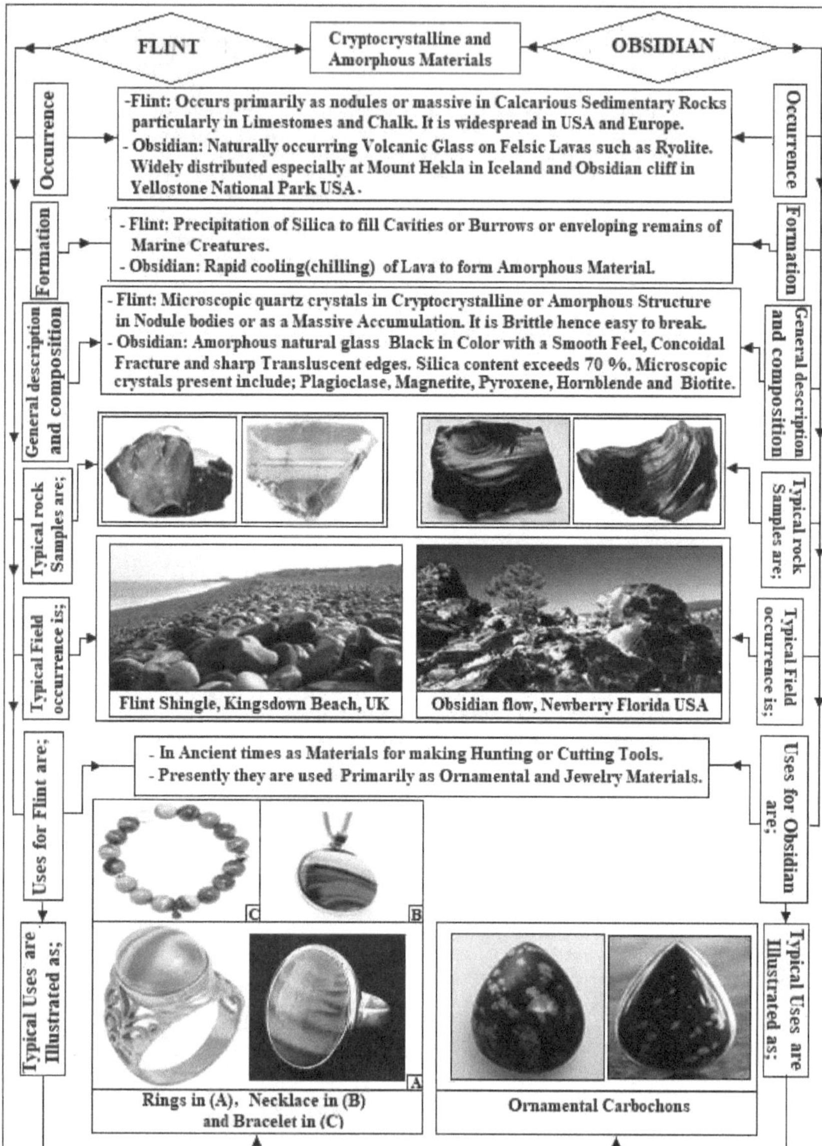
FLINT
Cryptocrystalline and Amorphous Materials
OBSIDIAN
Occurrence
-Flint: Occurs primarily as nodules or massive in Calcarious Sedimentary Rocks particularly in Limestomes and Chalk. It is widespread in USA and Europe.
- Obsidian: Naturally occurring Volcanic Glass on Felsic Lavas such as Ryolite. Widely distributed especially at Mount Hekla in Iceland and Obsidian cliff in Yellostone National Park USA.
Occurrence
Formation
- Flint: Precipitation of Silica to fill Cavities or Burrows or enveloping remains of Marine Creatures.
- Obsidian: Rapid cooling(chilling) of Lava to form Amorphous Material.
Formation
General description and composition
- Flint: Microscopic quartz crystals in Cryptocrystalline or Amorphous Structure in Nodule bodies or as a Massive Accumulation. It is Brittle hence easy to break.
- Obsidian: Amorphous natural glass Black in Color with a Smooth Feel, Concoidal Fracture and sharp Transluscent edges. Silica content exceeds 70 %. Microscopic crystals present include; Plagioclase, Magnetite, Pyroxene, Hornblende and Biotite.
General description and composition
Typical rock Samples are;
Typical rock Samples are;
Typical Field occurrence is;
Flint Shingle, Kingsdown Beach, UK
Obsidian flow, Newberry Florida USA
Typical Field occurrence is;
Uses for Flint are;
- In Ancient times as Materials for making Hunting or Cutting Tools.
- Presently they are used Primarily as Ornamental and Jewelry Materials.
Uses for Obsidian are;
C
B
A
Typical Uses are Illustrated as;
Rings in (A), Necklace in (B) and Bracelet in (C)
Ornamental Carbochons
Typical Uses are Illustrated as;

CHAPTER TEN

EXERCISES FOR SENDIMENTARY ROCKS

1.0 **(a)** Distinguish the term 'weathering' from 'erosion' as applied to sedimentary rocks.

(b) Describe processes of freezing and thawing that are prevalent in high altitude regions capturing features of sediments that are generated.

2.0 Distinguish void features that are prevalent in limestones from those that are associated with volcanic rocks capturing their role in mineral crystallization.

3.0 **(a)** Explain the following processes as applied to chemical weathering while focusing on typical chemical reactions and weathering features that develop: (i) Solution (ii)Hydrolysis (iii) Oxidation

(b) Distinguish exfoliation from sheeting as applied to physical weathering.

4.0 Describe spheroidal form of weathering while considering the mineralogical composition for typical rock types that exhibit this type of weathering.

5.0 **(a)** Explain the following:

(i) Forms of weathering; (i) Biotabation (ii) microbial activities. (ii) Chemical and physical impacts caused by plant roots to rock masses.

(b) Explain the role of the following factors in weathering process: (i) Climate (ii) Rock Type (iii) Rock Structure (iv)Topography

(v) Erosion (vi) Geologic time

6.0 **(a)** Explain effects of weathering that is anticipated long joints of rock masses such as granite.

(b) Explain weathering processes that affect stream sediments during transportation from granitic source rocks to a beach environment.

7.0 **(a)** Outline a digenetic processes that transform carbonate sediments into a bed rock.

(b) Distinguish a sedimentary rock that is cemented by secondary quartz from a similar rock that is cemented by secondary calcite if subjected to stress.

8.0 Discuss possible effects of chemical weathering on rocks such as;

(a) Limestones **(b** Sandstones.

9.0 Compare physical and mineralogical characteristics between the following:

(a) Conglomerate and Agglomerate

(b) Silt sediments and gravel sediments.

10.0 (a) Discuss the compositional framework for sandstones.

(b) Distinguish quartz sandstone from feldspathic sandstone.

(c) Outline characteristics of Mudrocks.

11.0 (a) Explain the following:

(i) Formation of evaporite considering environmental factors of influence.

(ii) Symbiosis as applied to lichen (ii) Ripple marks

(b) Outline occurrences and uses of the following (i) Basalt (ii) Gypsum

(iii) Tuff (iv) Laterite (v) color in sedimentary rocks giving rock examples.

(c) Distinguish characteristic for Rock salt from Rock crystal.

12.0 (a) Distinguish the following: (i) anthracite from lignite (ii) Rounding from Sphericity (iii) Graded bedding from cross bedding

(b) Describe sedimentary activities that are associated with formation of;

(i) Flute casts (ii) Cross stratification structures.

13.0 Discuss the formation process for the following while capturing environmental conditions that existed: **(a)** Pyroclastics **(b)** Banded ironstones.

14.0 Distinguish 'Sorting' from 'Grading' as applied to sedimentary rocks.

15.0 A student proposes that increase in weathering of a rock mass leads to a decrease in density and strength as well as an increase in deformability. Discuss this proposal capturing role of moisture and mineral bonding.

16.0 An ASM prospector decides to search for gemstones in an abandoned underground mine of limestone deposit. Discus the possible gemstones that the prospector can encounter capturing formation process for the gemstones.

EXPANDED EXERCISES WITH QUESTIONS AND ANSWERS

1.0 (a) Distinguish the term 'weathering' from 'erosion' as applied to sedimentary rocks.

Question: What is the difference between weathering and erosion in the context of sedimentary rocks?

Answer: Weathering refers to the process of breaking down rocks and minerals at their original location through physical, chemical, or biological means. It involves the alteration of the rock's structure and composition without movement. Erosion, on the other hand, is the process by which weathered materials are removed from their original location and transported to new sites by agents such as water, wind, or ice. In summary, weathering is the breakdown of rocks, while erosion is the movement of those broken materials.

(b) Describe processes of freezing and thawing that are prevalent in high altitude regions capturing features of sediments that are generated.

Question: How do freezing and thawing processes affect sediment generation in high-altitude regions?

Answer: In high-altitude regions, the freeze-thaw cycle occurs when water infiltrates cracks in rocks and freezes at low temperatures. The expansion of ice as it freezes exerts pressure on the surrounding rock, causing it to fracture. When temperatures rise, the ice melts, and the water can seep deeper into the cracks. This cycle repeats, leading to the gradual breakdown of the rock into smaller fragments, known as frost wedging. The sediments generated from this process are typically angular and poorly sorted, as they are derived from the mechanical disintegration of the parent rock.

2.0 Distinguish void features that are prevalent in limestones from those that are associated with volcanic rocks capturing their role in mineral crystallization.

Question: What are the differences between void features in limestones and those in volcanic rocks, and how do they contribute to mineral crystallization?

Answer: In limestones, void features such as caves and karst formations are often created through chemical weathering processes, particularly dissolution by acidic water. These voids can lead to the formation of secondary minerals like calcite and aragonite as the dissolved minerals precipitate in the cavities. In contrast, volcanic rocks may exhibit voids known as vesicles, which are formed by gas bubbles trapped during the solidification of lava. These vesicles can also contribute to mineral crystallization when minerals precipitate from the gas or fluid that fills the voids. Thus, while both types of rocks can have voids, the processes and resulting mineral formations differ significantly.

3.0 (a) Explain the following processes as applied to chemical weathering while focusing on typical chemical reactions and weathering features that develop:

(i) **Solution**

Question: What is the solution process in chemical weathering, and what are its typical reactions?

Answer: The solution process involves the dissolution of minerals in water, particularly when the water is slightly acidic due to dissolved carbon dioxide forming carbonic acid. For example, the dissolution of calcite in limestone can be represented by the reaction:

$$CO_2 + H_2O = H_2CO_3$$

Carbon dioxide water Carbolic acid

Carbonic acid dissociates to produce hydrogen and bicarbonate ions as follows:

$$H_2CO_3 \Leftrightarrow H^+ + HCO_3^-$$

Carbolic acid hydrogen ion bicarbonate ion

Typical reaction for chemical weathering of silicate minerals by carbonic acid;

$2KAlSiO_3O_8$ + H2O + 2H2CO3 = $AlSi_2O_5(OH)_2 + 2H^+ + 4SiO_2 + 2$ HCO_3^-

Feldspar water carbonic acid clay minerals dissolved ions

CaCO3 (s) + H2CO3 (aq) Ca2+ (aq) + 2HCO3 (aq)]

This process leads to the formation of voids and karst features in limestone.

(ii) **Hydrolysis**

Question: How does hydrolysis contribute to chemical weathering, and what are the typical reactions involved?

Answer: Hydrolysis is a chemical weathering process where minerals react with water, leading to the breakdown of the mineral structure. A common example is the hydrolysis of feldspar to form clay minerals:

A typical mineral reaction i.e. hydrolysis that involves albite and water is in the form;

$4NaAlSi_3O_8$ + 6H2O = $Al_4Si_4O_{10}(OH)_8 + 8SiO_2 + 4Na^+ + 4OH^-$

Albite water kaolinite silica sodium ion hydroxyl ion

Thus, the albite weathers to form kaolinite and produces some residual silica. Similarly, alkali feldspar produces clay minerals (Kaolinite) as follows:

KAlSi3O8 + H2O => Al2Si2O5 (OH)4 +SiO2 + K^+

Feldspar water kaolinite silica potassium ions

This process results in the formation of clay minerals and contributes to soil development.

(iii) **Oxidation**

Question: What is oxidation in the context of chemical weathering, and what are its typical reactions?

1. **Answer:** Oxidation is a chemical weathering process where minerals, particularly those containing iron, react with oxygen to form oxides. For

example, the oxidation of iron-bearing minerals can be represented as: For example, iron-bearing minerals such as olivine, pyroxene, and biotite can react with oxygen to form iron oxides. This reaction can be represented as follows:

(a) $4Fe^{+2}$ $+3O_2$ = $2Fe_2O_3$

Ferrous iron (Fe^{+2}) oxygen Ferric iron (Fe^{+3})

e.g. in iron e.g. hematite

(b) Olivine weathers through the combination of hydrolysis and oxidation to form Hematite, Thus;

$$2Fe_2SiO_4 + 4H_2O + O_2 = 2Fe_2O_3 + 2H_4SiO_4$$

Olivine water oxygen hematite silicic acid

This process leads to the formation of rust-colored iron oxides, which can weaken the rock structure and contribute to its disintegration.

(b) Distinguish exfoliation from sheeting as applied to physical weathering.

Question: How do exfoliation and sheeting differ in the context of physical weathering?

Answer: Exfoliation is a physical weathering process where layers of rock peel away from the surface due to thermal expansion and contraction, often seen in granite. This results in the formation of curved, sheet-like layers. Sheeting, on the other hand, refers to the release of pressure from overlying materials, causing the rock to expand and fracture along parallel planes. While both processes result in the separation of rock layers, exfoliation is primarily driven by temperature changes, whereas sheeting is related to pressure release.

4.0 Describe spheroidal form of weathering while considering the mineralogical composition for typical rock types that exhibit this type of weathering.

Question: What is spheroidal weathering, and which rock types typically exhibit this form of weathering?

Answer: Spheroidal weathering is a process where angular rock masses are transformed into rounded shapes due to the preferential weathering of corners and edges. This occurs as water penetrates along joints and fractures, leading to the breakdown of minerals. Common rock types that exhibit spheroidal weathering include granite and other granitic rocks, where feldspar and quartz are the primary minerals. The process results in the formation of rounded boulders and can enhance the rock's porosity and permeability.

5.0 (a) Explain the following:

(i) **Forms of Weathering:**

- **Bioturbation**: Bioturbation refers to the disturbance of soil and sediment by living organisms, such as earthworms and insects. This process can enhance soil aeration, mixing, and nutrient cycling, contributing to weathering.
- **Microbial Activities**: Microbial activities involve the breakdown of minerals and organic matter by microorganisms. These activities can lead to chemical weathering through processes such as organic acid production, which facilitates mineral dissolution.

(ii) **Chemical and Physical Impacts Caused by Plant Roots to Rock Masses**:

Question: How do plant roots impact rock masses through chemical and physical impacts?

Answer: Plant roots can significantly affect rock masses through both chemical and physical processes:

- **Chemical Impacts**: As roots grow, they exude organic acids that can chemically weather minerals in the surrounding rock. For example, the acids can dissolve calcium carbonate in limestone or facilitate the hydrolysis of feldspar in granite, leading to the formation of clay minerals. This chemical weathering process not only alters the mineral composition of the rock but also contributes to soil formation and nutrient availability.
- **Physical Impacts**: The physical presence of plant roots can exert pressure on rock masses as they grow and expand. This pressure can lead to the fracturing of rocks, particularly along existing joints and cracks. Over time, the mechanical action of roots can contribute to the disintegration of rock material, enhancing the overall weathering process.

(b) Explain the role of the following factors in the weathering process:

(i) **Climate**: Climate plays a crucial role in weathering processes. In warm, humid climates, chemical weathering predominates due to higher temperatures and moisture levels, which accelerate chemical reactions. Conversely, in cold or arid climates, physical weathering processes, such as freeze-thaw cycles, are more prevalent.

(ii) **Rock Type**: Different rock types exhibit varying susceptibility to weathering based on their mineral composition and structure. For example, granite, composed of quartz and feldspar, weathers differently than limestone, which is primarily made of calcite. The mineralogical composition influences the rates and types of weathering processes that occur.

(iii) **Rock Structure**: The structural features of rocks, such as joints, fractures, and bedding planes, can significantly influence weathering. Rocks with numerous joints are more susceptible to weathering because water and other weathering agents can penetrate more easily, enhancing both physical and chemical weathering processes.

(iv) **Topography**: The landscape's topography affects drainage patterns and microclimates, which in turn influence weathering. Steep slopes may experience more erosion and physical weathering, while flatter areas may accumulate moisture, promoting chemical weathering.

(v) **Erosion**: Erosion is closely linked to weathering, as it removes weathered materials from their original location. The removal of weathered material can expose fresh rock surfaces to further weathering processes, creating a feedback loop that accelerates the overall weathering process.

(vi) **Geologic Time**: The duration of exposure to weathering agents is critical. Over long periods, even resistant rocks can be significantly altered by weathering processes. Geologic time allows for the accumulation of weathering products, soil formation, and the development of distinct landforms.

6.0 (a) Explain the effects of weathering that are anticipated along joints of rock masses such as granite.

Question: What effects of weathering can be expected along the joints of granite rock masses?

Answer: Weathering along the joints of granite can lead to several significant effects:

- **Increased Fracturing**: Weathering processes, particularly chemical weathering, can weaken the rock along the joints, leading to further fracturing and disintegration. This can create a network of cracks that enhances the rock's susceptibility to erosion.
- **Formation of Clay Minerals**: The hydrolysis of feldspar in granite can lead to the formation of clay minerals, which can fill the joints and contribute to soil development.
- **Spheroidal Weathering**: The presence of joints allows for the penetration of water, which can lead to spheroidal weathering. This process results in the rounding of rock masses as corners and edges are preferentially weathered.
- **Increased Porosity and Permeability**: As weathering progresses along the joints, the rock's porosity and permeability can increase, allowing for greater water infiltration and further weathering processes.

(b) Explain the weathering processes that affect stream sediments during transportation from granitic source rocks to a beach environment.

Question: What weathering processes affect stream sediments as they are transported from granitic source rocks to a beach environment?

Answer: Several weathering processes can affect stream sediments during their transportation:

- **Mechanical Weathering**: As sediments are transported by the stream, they undergo mechanical weathering through abrasion against each other and the streambed. This process can lead to the rounding of sediment particles and the reduction of their size.
- **Chemical Weathering**: During transport, sediments may be exposed to water and dissolved gases, leading to chemical weathering. For example, feldspar grains may undergo hydrolysis, resulting in the formation of clay minerals. Additionally, oxidation of iron-bearing minerals can occur, leading to color changes in the sediments.

- **Sorting**: As sediments are transported, they are sorted by size and density due to variations in water velocity. Finer particles may be carried further downstream, while coarser materials settle out first. This sorting process can lead to distinct sedimentary layers in the beach environment.
- **Deposition**: Upon reaching the beach, sediments are deposited, where they may undergo further weathering processes influenced by wave action, tidal movements, and exposure to atmospheric conditions.

7.0 (a) Outline diagenetic processes that transform carbonate sediments into bedrock.

Question: What diagenetic processes transform carbonate sediments into bedrock?

Answer: The transformation of carbonate sediments into bedrock involves several diagenetic processes:

- **Compaction**: As sediments accumulate, the weight of overlying materials compresses the lower layers, reducing pore space and expelling water.
- **Cementation**: Minerals precipitate from pore fluids and fill the spaces between sediment grains, binding them together. Common cements in carbonate rocks include calcite and aragonite.
- **Recrystallization**: During diagenesis, the original carbonate minerals may undergo recrystallization, where smaller crystals dissolve and larger crystals form, enhancing the rock's strength and stability.
- **Dolomitization**: In some cases, magnesium-rich fluids can replace calcium in calcite, transforming limestone into dolomite, resulting in a rock that has different physical and chemical properties.

(b) Distinguish a sedimentary rock that is cemented by secondary quartz from a similar rock that is cemented by secondary calcite if subjected to stress.

Question: How can you distinguish between a sedimentary rock cemented by secondary quartz and one cemented by secondary calcite when subjected to stress?

Answer: When subjected to stress, sedimentary rocks cemented by secondary quartz and those cemented by secondary calcite exhibit different behaviors:

- **Quartz-Cemented Rocks**: Rocks cemented with secondary quartz tend to be more resistant to stress and exhibit greater hardness. When subjected to stress, they may show minimal deformation and can fracture in a brittle manner. The fractures typically occur along the grain boundaries, and the rock may produce a sharp, clean break.

- **Calcite-Cemented Rocks**: In contrast, rocks cemented with secondary calcite are generally softer and more susceptible to stress. When subjected to stress, they may deform more easily and can exhibit ductile behavior, leading to bending or warping before fracturing. The fractures in calcite-cemented rocks may be more irregular and can produce a more granular appearance due to the softer nature of calcite compared to quartz.

8.0 Discuss possible effects of chemical weathering on rocks such as:

(a) **Limestones**

Question: What are the possible effects of chemical weathering on limestone?

Answer: Chemical weathering significantly affects limestone in several ways:

- **Dissolution**: Limestone, primarily composed of calcite, is highly susceptible to dissolution by acidic waters. This process can lead to the formation of karst landscapes, characterized by features such as sinkholes, caves, and underground rivers.

- **Formation of Secondary Minerals**: As limestone dissolves, it can lead to the precipitation of secondary minerals, such as calcite and aragonite, in voids and fractures, contributing to the development of speleothems (stalactites and stalagmites) in caves.

- **Soil Formation**: The dissolution of limestone contributes to soil formation in karst regions, as the released calcium ions can enhance soil fertility.

(b) **Sandstones**

Question: What are the possible effects of chemical weathering on sandstones?

Answer: Chemical weathering affects sandstones in various ways:

- **Alteration of Feldspar**: Many sandstones contain feldspar, which can undergo hydrolysis to form clay minerals. This process can lead to a reduction in the rock's strength and stability.

- **Cement Dissolution**: If the sandstone is cemented by calcite or other soluble minerals, chemical weathering can lead to the dissolution of the cement, resulting in increased porosity and reduced cohesion among the grains.
- **Color Changes**: Oxidation of iron-bearing minerals within the sandstone can lead to color changes, often resulting in reddish or yellowish hues due to the formation of iron oxides.

9.0 Compare physical and mineralogical characteristics between the following:

(a) **Conglomerate and Agglomerate**

Question: What are the physical and mineralogical differences between conglomerate and agglomerate?

Answer:

- **Conglomerate**:
  - **Physical Characteristics**: Conglomerate is a clastic sedimentary rock composed of rounded gravel-sized clasts (greater than 2 mm in diameter) that are cemented together. It typically has a coarse texture and may exhibit a variety of colors depending on the composition of the clasts.
  - **Mineralogical Characteristics**: The clasts in conglomerate can be composed of various rock types, including quartz, granite, and volcanic rocks, and are often well-rounded due to transport by water.
- **Agglomerate**:
  - **Physical Characteristics**: Agglomerate is a volcanic rock composed of larger, angular fragments (greater than 2 mm) that are typically ejected during explosive volcanic eruptions. It has a more chaotic texture and may contain a mix of sizes and shapes.
  - **Mineralogical Characteristics**: The fragments in agglomerate are primarily volcanic in origin, including pumice, basalt, and other volcanic materials, and are often less rounded than those found in conglomerate.

(b) **Silt Sediments and Gravel Sediments**

Question: What are the physical and mineralogical differences between silt sediments and gravel sediments?

Answer:

- **Silt Sediments**:
  - **Physical Characteristics**: Silt sediments consist of fine particles with a diameter between 0.002 mm and 0.063 mm. They feel smooth to the touch and can retain moisture well, leading to a cohesive texture when wet.
  - **Mineralogical Characteristics**: Silt is often composed of quartz, feldspar, and clay minerals, and it can contain organic material. The fine particle size allows for significant surface area, which can enhance chemical weathering processes.
- **Gravel Sediments**:
  - **Physical Characteristics**: Gravel sediments consist of larger particles, typically greater than 2 mm in diameter. They are coarse and can include a range of sizes from pebbles to boulders. Gravel is less cohesive than silt and drains well.
  - **Mineralogical Characteristics**: Gravel is often composed of a variety of rock types, including quartz, granite, and basalt. The larger particle size means that gravel is less affected by chemical weathering compared to finer sediments.

10.0 (a) Discuss the compositional framework for sandstones.

Question: What is the compositional framework for sandstones?

Answer: The compositional framework for sandstones typically includes three main components:

- **Framework Grains**: The primary constituents of sandstone, which are usually quartz grains, but can also include feldspar and lithic fragments. The proportion of these grains determines the classification of the sandstone (e.g., quartz sandstone, feldspathic sandstone).

- **Matrix**: The finer-grained material that fills the spaces between the framework grains. The matrix can consist of silt, clay, or other fine particles and can influence the rock's porosity and permeability.
- **Cement**: The mineral material that binds the framework grains together. Common cements include silica (quartz), calcite, and iron oxides. The type and amount of cement can significantly affect the rock's strength and durability.

(b) Distinguish quartz sandstone from feldspathic sandstone.

Question: How can you differentiate between quartz sandstone and feldspathic sandstone?

Answer:

- **Quartz Sandstone**:
    - **Composition**: Quartz sandstone is primarily composed of more than 90% quartz grains. It is characterized by its high resistance to weathering due to the durability of quartz.
    - **Physical Characteristics**: Quartz sandstone typically has a well-sorted texture, with grains that are rounded and uniform in size. The color can vary but is often light-colored, ranging from white to tan.
    - **Cement**: The cement in quartz sandstone is often silica, which further enhances its strength and resistance to weathering.
- **Feldspathic Sandstone**:
    - **Composition**: Feldspathic sandstone contains a significant amount of feldspar, typically comprising 5% to 25% of the rock. The presence of feldspar distinguishes it from quartz sandstone.
    - **Physical Characteristics**: Feldspathic sandstone may have a less uniform texture compared to quartz sandstone, with a mix of grain sizes and shapes. It can appear more colorful, often exhibiting shades of pink, red, or brown due to the feldspar content.
    - **Cement**: The cement in feldspathic sandstone can vary but may include clay minerals or calcite, which can affect its overall durability.

(c) Outline characteristics of mudrocks.

Question: What are the key characteristics of mudrocks?

Answer: Mudrocks are a category of sedimentary rocks that include both siltstone and shale. Their key characteristics include:

- **Grain Size**: Mudrocks are composed of fine-grained particles, primarily silt and clay, with grain sizes less than 0.063 mm. This fine texture gives mudrocks a smooth feel when wet.
- **Fissility**: Many mudrocks, particularly shale, exhibit fissility, which means they can easily split into thin layers or sheets along parallel planes. This property is due to the alignment of clay minerals during deposition.
- **Color**: Mudrocks can vary in color, often appearing gray, black, or brown, depending on the organic content and mineral composition. Darker colors may indicate higher organic material, while lighter colors may suggest a higher proportion of quartz or feldspar.
- **Porosity and Permeability**: Mudrocks generally have low porosity and permeability due to the fine-grained nature of the sediments, which can limit fluid movement. However, they can serve as effective seals for hydrocarbons in petroleum reservoirs.
- **Organic Content**: Mudrocks often contain organic material, which can contribute to the formation of fossil fuels, such as coal and oil shale.

11.0 (a) Explain the following:

(i) **Formation of Evaporite Considering Environmental Factors of Influence**

Question: How do environmental factors influence the formation of evaporites?

Answer: Evaporites form in arid or semi-arid environments where evaporation rates exceed precipitation. Key environmental factors influencing their formation include:

- **Climate**: Hot, dry climates promote high evaporation rates, leading to the concentration of dissolved minerals in water bodies such as lakes or seas.
- **Water Source**: Evaporites typically form in enclosed basins where water can accumulate and become supersaturated with minerals. The lack of outflow allows for continued evaporation and mineral precipitation.

- **Mineral Composition**: The specific minerals that precipitate as evaporites depend on the chemical composition of the water. Common evaporite minerals include halite (rock salt), gypsum, and anhydrite, which form in sequence as evaporation progresses.

(ii) **Symbiosis as Applied to Lichen**

Question: What is the significance of symbiosis in lichens?

Answer: Lichens are a symbiotic association between fungi and photosynthetic organisms, typically algae or cyanobacteria. The significance of this symbiosis includes:

- **Nutrient Exchange**: The photosynthetic partner provides organic carbon and energy through photosynthesis, while the fungal partner offers protection and access to moisture and minerals from the substrate.

- **Weathering Contribution**: Lichens contribute to weathering processes by secreting organic acids that can dissolve minerals in rocks, facilitating the breakdown of the substrate and contributing to soil formation.

(iii) **Ripple Marks**

Question: What are ripple marks, and how do they form?

Answer: Ripple marks are small, wave-like structures that form on the surface of sedimentary deposits, typically in environments with flowing water or wind. They are created by the movement of water or air over loose sediments, which causes the sediment to be displaced and shaped into ripples. Ripple marks can provide valuable information about the direction of current flow and the energy conditions of the depositional environment.

(b) Outline occurrences and uses of the following:

(i) **Basalt**

Question: Where is basalt commonly found, and what are its uses?

Answer: Basalt is a volcanic rock that is commonly found in oceanic crust and on land in volcanic regions. It is formed from the rapid cooling of lava. Occurrences include:

- **Oceanic Basins**: Basalt forms the majority of the ocean floor, creating mid-ocean ridges and volcanic islands.

- **Continental Flood Basalts**: Large volcanic provinces, such as the Columbia River Basalt Group in the United States, are composed of extensive basalt flows.

Uses: Basalt is used in construction (as aggregate for concrete), road base material, and in the production of stone wool insulation. It is also used in landscaping and as a decorative stone.

(ii) **Gypsum**

Question: Where is gypsum found, and what are its applications?

Answer: Gypsum is a sedimentary evaporite mineral commonly found in sedimentary rock formations, particularly in arid regions. It forms from the evaporation of seawater or saline lakes. Occurrences include:

- **Evaporite Deposits**: Gypsum is often found in large deposits associated with ancient marine environments, such as the Permian Basin in Texas.

Uses: Gypsum is widely used in the construction industry for making drywall (gypsum board), plaster, and cement. It is also used as a soil conditioner in agriculture to improve soil structure and drainage.

(iii) **Tuff**

Question: What is tuff, where is it found, and what are its uses?

Answer: Tuff is a type of volcanic rock formed from the consolidation of volcanic ash and debris ejected during explosive eruptions. Occurrences include:

- **Volcanic Regions**: Tuff is commonly found in areas with past volcanic activity, such as the Tertiary volcanic fields in the western United States and the Mediterranean region.

Uses: Tuff is used as a building material, particularly in regions where it is abundant. It can be carved easily and is often used for decorative stone and architectural applications. Additionally, tuff can be used as a lightweight aggregate in concrete and as a soil amendment in agriculture due to its mineral content.

(iv) **Laterite**

Question: What is laterite, where is it commonly found, and what are its uses?

Answer: Laterite is a type of soil and rock rich in iron and aluminum, formed in tropical and subtropical climates through intense weathering of parent rock. Occurrences include:

- **Tropical Regions**: Laterite is commonly found in regions with high rainfall and warm temperatures, such as parts of India, Southeast Asia, and Africa. It often forms in areas where the weathering process is accelerated due to the warm, humid climate.

Uses: Laterite is used as a building material, particularly in regions where it is abundant. It can be cut into bricks for construction. Additionally, laterite is used in the production of iron and aluminum ores and as a soil amendment in agriculture to improve soil fertility.

(v) **Color in Sedimentary Rocks**

Question: How does color influence the identification and interpretation of sedimentary rocks?

Answer: The color of sedimentary rocks can provide valuable information about their mineral composition, depositional environment, and diagenetic history. For example:

- **Red and Yellow Colors**: Often indicate the presence of iron oxides, suggesting oxidation processes in well-drained environments.
- **Gray and Black Colors**: Typically indicate organic-rich sediments, such as those found in marine or swampy environments, where reduced conditions prevail.
- **White or Light Colors**: Often suggest the presence of quartz or carbonate minerals, indicating a more arid environment or the influence of chemical precipitation.

Understanding the color of sedimentary rocks can help geologists infer past environmental conditions and processes.

(c) Distinguish characteristics for Rock Salt from Rock Crystal.

Question: What are the distinguishing characteristics of rock salt compared to rock crystal?

Answer:

- **Rock Salt (Halite)**:
  - **Composition**: Rock salt is primarily composed of sodium chloride ($NaCl$).
  - **Physical Characteristics**: It typically appears as a crystalline, translucent to transparent rock with a cubic crystal structure. It has a characteristic salty taste and is soluble in water.
  - **Occurrence**: Rock salt forms from the evaporation of saline water in arid environments, often found in evaporite deposits.
- **Rock Crystal (Quartz)**:
  - **Composition**: Rock crystal is composed of silicon dioxide (SiO_2).
  - **Physical Characteristics**: It is usually clear or translucent and exhibits a hexagonal crystal structure. Rock crystal is harder than rock salt and does not have a salty taste.
  - **Occurrence**: Rock crystal can form in a variety of geological environments, including igneous, metamorphic, and sedimentary settings.

12.0 (a) Distinguish the following:

(i) **Anthracite from Lignite**

Question: What are the differences between anthracite and lignite?

Answer:

- **Anthracite**:
  - **Composition**: Anthracite is a high-grade coal with a carbon content of over 86%. It has a shiny appearance and is hard and brittle.
  - **Properties**: It burns with a clean flame and produces a high amount of heat, making it a desirable fuel source. Anthracite has low volatile matter and is often used for heating and in industrial processes.

- **Lignite**:
  - **Composition**: Lignite, also known as brown coal, has a lower carbon content (about 25-35%) and higher moisture content compared to anthracite.
  - **Properties**: It is soft, crumbly, and has a brownish color. Lignite burns with a smoky flame and produces less heat than anthracite. It is primarily used for electricity generation in power plants.

(ii) **Rounding from Sphericity**

Question: How do rounding and sphericity differ in sedimentary particles?

Answer:

- **Rounding**: Rounding refers to the degree to which the edges and corners of sedimentary particles have been smoothed and worn down due to abrasion during transport. Well-rounded particles have smooth, curved edges, while angular particles retain sharp edges.
- **Sphericity**: Sphericity is a measure of how closely the shape of a particle resembles a perfect sphere. It considers the dimensions of the particle in three dimensions. A particle with high sphericity is more spherical in shape, while one with low sphericity may be elongated or irregular.

(iii) **Graded Bedding from Cross Bedding**

Question: What are the differences between graded bedding and cross bedding?

Answer:

- **Graded Bedding**: Graded bedding is characterized by a vertical change in grain size within a single layer, typically showing a coarser material at the bottom that gradually fines upward. This feature often indicates a decrease in energy conditions during deposition, such as in a turbidity current.
- **Cross Bedding**: Cross bedding consists of inclined layers within a sedimentary bed, formed by the migration of ripples or dunes. The angles of the cross beds indicate the direction of sediment transport and are typically found in environments with consistent current flow, such as rivers or wind-blown dunes.

13.0 Discuss the formation process for the following while capturing environmental conditions that existed:

(a) **Pyroclastics**

Question: What is the formation process of pyroclastics, and what environmental conditions contribute to their development?

Answer: Pyroclastics are formed from volcanic eruptions that explosively eject fragments of molten rock, ash, and gases into the atmosphere. The formation process includes:

- **Eruption**: During a volcanic eruption, magma rises to the surface and can be explosively released, fragmenting into various sizes of particles, from fine ash to larger volcanic bombs.
- **Cooling and Solidification**: As these particles are ejected into the air, they cool rapidly and solidify, forming pyroclastic materials. The size and composition of the pyroclastic deposits depend on the eruption's intensity and the magma's viscosity.
- **Environmental Conditions: Pyroclastic deposits typically form in environments characterized by explosive volcanic activity. These conditions can include:**
- **Subduction Zones**: Many pyroclastic deposits are associated with stratovolcanoes found at convergent plate boundaries, where an oceanic plate subducts beneath a continental plate. The intense pressure and heat lead to the generation of volatile-rich magma, which can result in explosive eruptions.
- **Hotspots**: Some pyroclastic materials can also form from hotspot volcanism, where magma rises from deep within the Earth, leading to explosive eruptions, as seen in places like the Hawaiian Islands.
- **Calderas**: Large volcanic calderas can produce extensive pyroclastic flows during massive eruptions, where the collapse of the volcano allows for the rapid ejection of volcanic material.

(b) **Banded Ironstones**

Question: What is the formation process of banded ironstones, and what environmental conditions contribute to their development?

Answer: Banded ironstones are sedimentary rocks characterized by alternating layers of iron-rich minerals and silica or chert. Their formation process includes:

- **Chemical Precipitation**: Banded ironstones form primarily through the precipitation of iron oxides (such as hematite and magnetite) from seawater. This process is often facilitated by the activity of photosynthetic microorganisms, which produce oxygen and lead to the oxidation of dissolved iron.
- **Environmental Conditions**: The formation of banded ironstones typically occurs in shallow marine environments, particularly during periods of high biological productivity. Key conditions include:
  - **Anoxic Conditions**: During periods when the ocean was anoxic (lacking oxygen), dissolved iron could accumulate in the water. When oxygen levels increased, iron would precipitate out of solution, forming iron-rich layers.
  - **Tectonic Activity**: Changes in sea level and tectonic uplift can create conditions conducive to the deposition of iron-rich sediments, leading to the formation of banded iron formations (BIFs).

14.0 Distinguish 'Sorting' from 'Grading' as applied to sedimentary rocks.

Question: What is the difference between sorting and grading in sedimentary rocks?

Answer:

- **Sorting**: Sorting refers to the distribution of grain sizes within a sedimentary rock or sediment deposit. Well-sorted sediments consist of grains that are similar in size, while poorly sorted sediments contain a mix of grain sizes. Sorting is influenced by the energy of the transporting medium; higher energy environments tend to produce well-sorted sediments, while lower energy environments can lead to poorly sorted deposits.
- **Grading**: Grading refers to the vertical change in grain size within a single sedimentary layer. Graded bedding typically shows a coarser material at the bottom that fines upward, indicating a decrease in energy conditions during deposition. This feature is often associated with processes such as turbidity currents, where sediment is deposited in a sequence based on particle size.

15.0 A student proposes that an increase in weathering of a rock mass leads to a decrease in density and strength as well as an increase in deformability. Discuss this proposal capturing the role of moisture and mineral bonding.

Question: How does weathering affect the density, strength, and deformability of rock masses?

Answer: The proposal that increased weathering leads to a decrease in density and strength, as well as an increase in deformability, is supported by several factors:

- **Moisture**: Weathering processes often involve the infiltration of water into rock masses. The presence of moisture can enhance chemical weathering reactions, leading to the breakdown of minerals and the formation of clay minerals. As minerals are altered or dissolved, the overall density of the rock mass decreases.
- **Mineral Bonding**: Weathering can weaken the bonds between mineral grains within a rock. For example, the dissolution of cementing materials (such as calcite or silica) can reduce the cohesion between grains, leading to a decrease in the rock's strength. As the rock becomes more porous and less cohesive, it becomes more susceptible to deformation under stress.
- **Deformability**: As weathering progresses, the rock mass may become more deformable due to the increased presence of clay minerals, which can absorb water and expand. This change in mineralogy can lead to a rock that is more pliable and less resistant to deformation.

Overall, the proposal highlights the interconnectedness of weathering processes, moisture, and mineral bonding in influencing the physical properties of rock masses.

16.0 A student proposes that an ASM prospector decides to search for gemstones in an abandoned underground mine of limestone deposit. Discuss the possible gemstones that the prospector can encounter capturing the formation process for the gemstones.

Question: What gemstones might a prospector find in an abandoned underground limestone mine, and what are their formation processes?

Answer: In an abandoned underground limestone mine, a prospector may encounter several types of gemstones, including:

- **Calcite**: While not a traditional gemstone, calcite can form beautiful crystal formations in limestone caves. Calcite crystals can be clear, white, or colored, and they form through the precipitation of calcium carbonate from mineral-rich water.
- **Fluorite**: Often found in limestone deposits, fluorite can occur in a variety of colors, including purple, green, and blue. It forms from the hydrothermal alteration of limestone and can crystallize in cubic forms. Fluorite is valued for its vibrant colors and is used in jewelry and decorative items.
- **Malachite and Azurite**: These copper carbonate minerals can form in the oxidation zones of copper deposits associated with limestone. Malachite is green, while azurite is blue, and both can form beautiful banded patterns. They are often used as gemstones and in ornamental carvings.
- **Opal**: In some limestone environments, opal can form as a result of the silica-rich waters that permeate the rock. Opal can exhibit a play of color and is highly valued in jewelry.
- **Onyx**: Onyx is a banded variety of chalcedony, which can form in limestone environments through the deposition of silica from groundwater. It is characterized by its parallel bands of color, typically black and white, and is often used in jewelry and decorative items.
- **Agate**: Similar to onyx, agate is another form of chalcedony that can form in limestone deposits. It is known for its colorful banding and can be found in a variety of colors. Agate forms through the slow deposition of silica in cavities within the limestone, often resulting in intricate patterns.
- **Sodalite**: While not as common, sodalite can sometimes be found in limestone deposits. It is a blue mineral that can form in metamorphosed limestone and is valued for its vibrant color.

Formation Processes for Gemstones

The formation processes for gemstones found in limestone deposits are diverse and intricate, reflecting the complex interplay of geological, chemical, and environmental factors. The following processes are key to the formation of these valuable minerals:

1. **Precipitation**: Many gemstones originate from the precipitation of minerals from mineral-rich water. This process often occurs in

environments where water becomes supersaturated with dissolved minerals. As conditions change—such as through evaporation, temperature fluctuations, or pressure changes—these dissolved minerals can crystallize. For example, in limestone environments, calcium carbonate can precipitate out of solution to form calcite, which can sometimes develop into beautiful crystalline structures. Other minerals, such as aragonite, can also form through similar precipitation processes, leading to the creation of gemstones like calcite crystals that exhibit unique colors and patterns.

2. **Hydrothermal Activity**: Hydrothermal processes play a significant role in the formation of certain gemstones, particularly those that require high temperatures and pressures. In limestone deposits, hydrothermal solutions—hot, mineral-rich fluids—can circulate through fractures and cavities within the rock. As these fluids cool or react with the surrounding rock, they can deposit minerals that crystallize into gemstones. For instance, fluorite can form in this manner, often exhibiting vibrant colors and distinct cubic crystal shapes. The presence of other elements in the hydrothermal fluids can also lead to the formation of gemstones with unique properties, such as the presence of trace elements that impart specific colors to the crystals.

3. **Metamorphism**: Metamorphic processes can also contribute to gemstone formation, particularly in limestone that is subjected to heat and pressure over geological time. When limestone undergoes metamorphism, it can transform into marble, a process that can alter the mineral composition and create new minerals. For example, the recrystallization of calcite during metamorphism can lead to the formation of high-quality calcite crystals, which may be considered gemstones. Additionally, the metamorphic environment can facilitate the growth of other minerals, such as dolomite, which can also crystallize into valuable gemstones. The conditions of metamorphism, including temperature and pressure, play a crucial role in determining the types of gemstones that can form.

4. **Oxidation and Weathering**: The weathering of limestone can lead to the formation of secondary minerals, particularly through the oxidation of copper-bearing minerals. In oxidized zones of limestone deposits, minerals such as malachite and azurite can form as a result of chemical reactions involving copper. These vibrant green and blue minerals are not only aesthetically pleasing but also valuable as gemstones. The weathering process can expose these minerals at the surface, making them accessible

for prospectors. Additionally, the alteration of primary minerals during weathering can create a variety of secondary minerals, contributing to the diversity of gemstones found in limestone environments.

Conclusion

In summary, an artisanal small-scale miner (ASM) searching for gemstones in an abandoned underground limestone mine may encounter a variety of valuable minerals, each with unique formation processes. The geological history of the limestone, combined with the effects of hydrothermal activity, precipitation, metamorphism, and weathering, contributes to the diversity of gemstones found in these environments. Understanding these processes not only aids in gemstone identification but also enhances the appreciation of the natural beauty and complexity of mineral formations. By recognizing the intricate relationships between geological processes and gemstone formation, prospectors can better navigate their search for these precious materials, ultimately leading to a deeper understanding of the Earth's geological history and the treasures it holds.

REFERENCES

Berry, LG., Mason, B. and Dietrich, R.V. (1983). Mineralogy: Concepts, Descriptions and Determinations. W.H. Freeman and Company, San Francisco, 561pp.

Billings, M. P (2016). Structural Geology, Published by Pearson Education Dorling, 3rd Edition, Kindersley, ISBN-13: 978-9332577565.

Bloss, F.D. (1971). Crystallography and Crystal Chemistry, An Introduction. New York: Holt,Rinehart & Winston, 545 pp.

Brannon, C.H. (2016). A Look at Erosion and Weathering: The Rock Cycle, 32 p, Published by Enslow Pub Inc., ISBN-13: 978-0766072800.

Brian, J and Noel, P. J. (2015). Origin of Carbonate Sedimentary Rocks, 1st Edition, 464 p, Published by American Geophysical Union, ISBN-13: 978-1118652732.

Clement, C.R., Skinner, E.M.W. (1985). A textural-genetic classification of kimberlites. *Trans. Geol.Soc. S. Afr.* 88, 403-409.

DuBois, C and Walsh, J. (1970). Minerals of Kenya. *Geological Survey of Kenya Bulletin no.* 11.

Fossen, H.(2016). Structural Geology, 2nd Edition, 524 p, Published by Cambridge University Press, ISBN-13: 978-1107057647.

Gill, R. (2012). Igneous Rocks and Processes: A Practical Guide, 438 p, Published by Wiley- Blackwell, ISBN-13: 978-0632063772.

Htoo, K.M, Htoon, S and Khin, T (2004): Study on Physical Properties and Chemical Composition of Some Myanmar Gems, *The Jour. Myan. Acad. Arts & Sc. 2004* Vol. II (Part One) No. 2 (Science) 175

Hollocher, K. (2014). A Pictorial Guide to Metamorphic Rocks in the Field, 1st Edition, 326 p, Published by CRC Press, ISBN-13: 978-1138026308.

Hobbs, B. E and Ord A. (2014). Structural Geology: The Mechanics of Deforming Metamorphic, 1st Edition, 680 p, Published by Elsevier, ISBN-13: 978-0124078208.

Karen L. W and Simmons, W. (2008). Pegmatite genesis: state of the art, *Eur. Journal, Mineral, Symbosium paper.* Porto,Publ. 2008, 20, pp. 421–438

Kruhl, J. H. (2017). Drawing Geological Structures: Geological Field Guide, 1st Edition, 232 p, Published by Wiley-Blackwell, ISBN-13: 978-1405182324

Mitchell, R. H. (2013). Kimberlites: mineralogy, geochemistry, and petrology. Springer Science & Business Media.

Nyanai, C.M. (2014): SGL 101: Materials of The Earth, Lecture 2

https://profiles.uonbi.ac.ke/cnyamai/classes/sgl-101-materials-earth

Pell, J. (1998). Kimberlite-hosted Diamonds. In Geological Fieldwork (1997), British Columbia Ministry of Employment and Investment, Paper 1998-1, pages 24L-1 to 24L-4.

Prothero, D. R. and Schwab, F. (2013). Sedimentary Geology, 3rd Edition, 593 p, Published by W. H. Freeman, ISBN-13: 978-1429231558.

Rop, B. K and Namwiba, W.H. (2018). Fundamentals of Applied Geology: Competency and Evaluation Approach, 345p. Verlag, Publisher: LAP LAMBERT Academic Publishing ISBN 978-613-9-57896-2.

Rop, B. K and Namwiba, W.H and Maina, C.W. (2019): Introduction to Geology 129p. Verlag, Publisher: LAP LAMBERT Academic Publishing, ISBN 978-613-9-82770-1

Rop, B. K., Digne, E.R and Namwiba W.H (2020). Applied Geology in Mining and Construction Practice: A Competency and Geohazard Mitigation Platform 213p.Verlag, Publisher: Scholar's Press ISBN 978-613-8-93406-6.

Rop, B. K., Seroni, A and Krop, I. (2020). Economic Impacts of ASM in Taita Taveta County, Kenya: Gemstones Mining 129p. Verlag/Publisher: LAP LAMBERT Academic Publishing, ISBN 978-620-2-51546-7.

Rop, B. K., Namwiba, W. H., & Keter, M. C. (2024). *Economic potential of Kericho Ainamoi bauxite: A geological evaluation approach report* (80 p.). LAP Lambert Academic Publishing.

Rop, B. K., Mayik, D. A., & Gach, G. K. (2023). *Petroleum source rocks characterisation: LT-1 well Northern Kenya*. LAP Lambert Academic Publishing. ISBN 978-620-6-15177-7.

Rop, B. K., Rwatangabo, D. E. R., & Namwiba, H. W. (2023). *ASM geological rocks and minerals uses: An illustrative comparative lab manual* (260 p.). LAP Lambert Academic Publishing. ISBN 978-621-6-15178-3.

Rop, B. K., & Namwiba, W. H. (2023). *Fundamentals of soil mechanics: A competency and evaluation based class lecture manual* (223 p.). Scholar's Press. ISBN 978-620-5-52401-5.

Rop, B. K., & Namwiba, W. H. (2023). *Laboratory manual in soil mechanics: Essential procedures for testing and reporting* (112 p.). Scholar's Press. ISBN-10: 6205524058; ISBN-13: 978-6205524053.

Rop, B. K., Namwiba, W. H., & Keter, M. C. (2023). *Geology, geophysics and exploration geochemistry: A synopsis perspective* (144 p.). ISBN-10: 6206165809; ISBN-13: 978-6206165804.

Rop, B. K., & Namwiba, W. H. (2022). *Applied geology and evaluation perspective* (368 p.). Scholar's Press. ISBN 978-613-8-97461-1.

Rop, B. K., Namwiba, W. H., & Keter, M. C. (2023). *Applied geology and geophysics in exploration: A synoptic perspective* (128 p.). ISBN-10: 6205523027; ISBN-13: 978-6205523025.

Rop, B. K., Rwatangabo, D. E. R., & Namwiba, W. H. (2023). *Properties, occurrences and uses of minerals and rocks: A comparative illustrative approach* (257 p.). Scholar's Press. ISBN 978-620-5-52235-6.

Rop, B. K., Mayik, D. A., & Gach, K. (2023). *Petroleum source rocks characterisation: LT-1 well Northern Kenya*. LAP Lambert Academic Publishing. ISBN 978-620-6-15177-7.

Rop, B. K., Wangari, J. M., & Namwiba, W. H. (2022). *Fundamentals of laboratory practices in mineralogy and petrology: Mineralogy and petrology (A handbook for comparative studies)* (216 p.). Scholar's Press. ISBN 978-613-8-96862-7.

Rop, B. K. (2021). *Petroleum perspective of the NW-Kenya basins: Evidence and emerging oil exploration issues*. LAP Lambert Academic Publishing. ISBN 978-620-3-86294-2.

Rop, B. K., & Kipkiror, L. J. (2021). *Marakwet East Lagam escarpment landslides: Mitigation measures, Kenya*. Scholar's Press. ISBN 978-3-639-66207-8.

Rop, B. K., Fedha, I. M., & Namwiba, W. H. (2021). *Workshop and laboratory practices: A competency and evaluation based guide* (96 p.). Scholar's Press. ISBN 978-613-8-95047-9.

Rop, B. K., Ketter, M. C., & Sawe, S. J. (2021). *Review of oil and gas prospectivity: Kenya Rift basins* (104 p.). Scholar's Press. ISBN 978-613-8-94692-2.

Rop, B. K., Nyamai, C. M., & Namwiba, W. H. (2020). *Narrative glossary of fundamentals for applied geology: A construction perspective* (117 p.). Scholar's Press. ISBN 978-613-8-93606-0.

Rop, B. K., Anyona, S., Kipseba, E. K., Namwiba, W. H., & Kipkiror, L. J. (2020). Review of geothermal resources as manifestations of volcanism in Eastern Africa. In *Proceedings of the Sustainable Research and Innovation Conference* (pp. 79-88). JKUAT Main Campus, Kenya. ISSN 2079-6226.

Rop, B. K., Kipkiror, L. J., & Kipseba, E. K. (2020). Geoenviron tourism sites in Northern Kenya Rift Valley: The case of Lake Bogoria, Lake Baringo, Kerio Valley, and Silali. In *Proceedings of the Sustainable Research and Innovation Conference* (pp. 126-135). JKUAT Main Campus, Kenya. ISSN 2079-6226.

Rop, B. K., Mwendwa, J. M., Maina, C., & Namwiba, W. H. (2019). Applied geology in mining engineering. In *Proceedings of the Sustainable Research and Innovation (SRI) International Conference* (pp. 1-10). JKUAT College of Engineering and Technology (CoETEC), Juja, Kenya.

Rop, B. K., Mwendwa, J. M., Maina, C., & Namwiba, W. H. (2019). Importance of applied geology in mining perspective. In *Proceedings of the 2nd Annual International Conference* (pp. 1-10). Machakos University, Machakos, Kenya.

Kiamba, E., Kasomo, R. M., Rop, B. K., & Mwendwa, J. M. (2019). Limestone production optimization by fragmentation control – Case study of Bissel Quarry. In *Proceedings of the 2nd Annual International Conference* (pp. 1-10). Machakos University, Machakos, Kenya.

Rop, B. K. (2021). *Petroleum perspective of the NW-Kenya basins: Evidence and emerging oil exploration issues.* Saarbrücken, Germany: LAP LAMBERT Academic Publishing. ISBN 978-620-3-86294-2.

Sandra, E. and Robles, C. (2008): Kimberlites associated with the Lucapa structure,Angola *PhD. Dissertion* Universitat De Barcelona, Spain

Strekeisen,A(2020). Kimberlite, *Article* https://www.alexstrekeisen.it/english/vulc/kimberlites.php

Theissen, K. and Swanson, J. (2014). Metamorphic Rocks: Rocks and Minerals, 48 p, Published by Core Library, ISBN-13: 978-1624033889.

Yoder, H. S. (2016). Evolution of the Igneous Rocks: Fiftieth Anniversary Perspectives, 602 p, Published by Princeton University Press, ISBN-13: 978-0691631554.

ABOUT THE BOOK

Sedimentary Rocks: A Comparative Illustrative Approach

The study of sedimentary rocks is fundamental to understanding the Earth's processes, resources, and geological history. "Sedimentary Rocks: A Comparative Illustrative Approach" serves as a comprehensive resource tailored for students, educators, and practitioners in the field of geology, particularly those engaged at the Certificate and Diploma levels. This book also aims to empower Artisanal Small-scale Miners (ASM) by equipping them with essential geological knowledge that enhances their skills and understanding of the geological aspects pertinent to their work.

Overview of Content

In this book, we delve into the fundamental concepts surrounding sedimentary rocks, including their formation, classification, and the various processes that shape them. Each chapter is meticulously structured to build upon the previous one, guiding readers through the complexities of weathering, erosion, transportation, and diagenesis. This progressive approach ensures that learners can grasp intricate concepts in a logical and coherent manner.

The material is presented in a clear and engaging style, utilizing a competency-based approach that aligns with UNESCO's ASM Certificate and Diploma Programmes. This alignment not only enhances the educational value of the book but also ensures that the content is relevant and applicable to the needs of ASM practitioners.

Visual Learning through Comparative Illustrations

A distinctive feature of this book is the integration of comparative illustrations, which juxtapose schematic drawings with real-world field images. This visual approach significantly enhances comprehension by bridging the gap between theoretical knowledge and practical application. By providing visual representations of sedimentary processes, readers can better appreciate the dynamic interactions that shape these rocks and their environments.

The illustrations serve not only as educational tools but also as a means to inspire curiosity and foster a deeper understanding of the geological processes at play.

This dual approach of combining schematic and photographic representations allows for a more holistic learning experience.

Engaging with the Material

To enrich the learning experience, we have incorporated a variety of questions and exercises throughout the text. These are designed to stimulate critical thinking and encourage readers to actively engage with the material. By assessing their understanding through these interactive elements, learners can reinforce their knowledge and apply what they have learned in practical contexts.

Acknowledgments

We acknowledge the contributions of numerous individuals in the development of this book. The insights and research of various authors have been invaluable in shaping our discussions and illustrations. We extend our heartfelt gratitude to these contributors, as well as to our colleagues and friends who have supported us throughout this journey, particularly during the challenges posed by the Covid-19 pandemic. Their encouragement and expertise have been instrumental in bringing this project to fruition.

Conclusion

As you embark on this exploration of sedimentary rocks, we hope that this book serves as a valuable resource that enhances your knowledge and inspires a greater appreciation for the dynamic processes that shape our planet. Whether you are a student, educator, or practitioner in the mining and construction industry, we believe that the insights contained within these pages will be beneficial in your endeavors.

We invite you to delve into the fascinating dynamic systems and processes that sedimentary rocks reveal about the Earth's history and the resources they provide for our future. Through this journey, may you gain a deeper understanding of the geological forces that have shaped our world and continue to influence our lives today.

The Importance of Sedimentary Rocks

Sedimentary rocks are not merely geological formations; they are vital indicators of the Earth's history and processes. They provide insights into past environments, climate changes, and biological evolution. Understanding sedimentary rocks is crucial for several reasons:

1. **Resource Exploration**: Sedimentary rocks are often reservoirs for essential natural resources, including fossil fuels, groundwater, and minerals. Knowledge of sedimentary processes aids in the exploration and sustainable extraction of these resources, which is particularly relevant for ASM practitioners who rely on these materials for their livelihoods.

2. **Environmental Indicators**: The composition and structure of sedimentary rocks can reveal significant information about past environmental conditions. For instance, the presence of certain fossils can indicate the climate and ecological conditions of the time when the rocks were formed. This information is invaluable for understanding current environmental changes and predicting future trends.

3. **Geological History**: Sedimentary rocks serve as a historical record of the Earth's geological past. By studying these rocks, geologists can reconstruct ancient landscapes, understand tectonic movements, and trace the evolution of life on Earth. This historical perspective is essential for comprehending the processes that have shaped our planet over millions of years.

4. **Soil Formation and Agriculture**: Sedimentary rocks contribute to soil formation, which is critical for agriculture. Understanding the mineral composition and weathering processes of these rocks can help in managing soil health and fertility, thereby supporting sustainable agricultural practices.

Target Audience

This book is designed for a diverse audience, including:

- **Students**: Those pursuing studies in geology, environmental science, and related fields will find this book a valuable resource for understanding sedimentary rocks and their significance.

- **Educators**: Instructors can utilize this text as a teaching aid, providing a structured approach to sedimentary geology that is both informative and engaging.

- **Practitioners**: Professionals in the mining and construction industries, particularly those involved in ASM, will benefit from the practical knowledge and skills presented in this book, enhancing their operational effectiveness and safety.

- **Researchers**: Scholars and researchers interested in sedimentology and related fields will find the comprehensive discussions and illustrations useful for their studies and projects.

Pedagogical Approach

The pedagogical approach adopted in this book emphasizes active learning and critical thinking. By incorporating questions, exercises, and comparative illustrations, we aim to foster an interactive learning environment. This approach not only aids in knowledge retention but also encourages learners to apply their understanding in real-world contexts.

The competency-based framework aligns with contemporary educational standards, ensuring that learners acquire practical skills alongside theoretical knowledge. This alignment is particularly beneficial for ASM practitioners, who often require a blend of academic understanding and hands-on experience to navigate the complexities of their work.

Future Directions

As we look to the future, the importance of sedimentary rocks in addressing global challenges cannot be overstated. Issues such as climate change, resource depletion, and environmental degradation necessitate a deeper understanding of geological processes. This book aims to contribute to that understanding by providing a solid foundation in sedimentary geology.

We encourage readers to continue exploring the dynamic systems and processes that govern our planet. By fostering a greater appreciation for sedimentary rocks, we hope to inspire future generations of geologists, environmental scientists, and resource managers to engage with the Earth's history and its resources responsibly and sustainably.

Final Thoughts

In conclusion, "Sedimentary Rocks: A Comparative Illustrative Approach" is more than just a textbook; it is a gateway to understanding the intricate relationships between geological processes, natural resources, and environmental stewardship. We invite you to embark on this educational journey, equipped with the knowledge and insights that will empower you to make informed decisions in your academic and professional pursuits.

As you turn the pages of this book, may you find inspiration in the stories that sedimentary rocks tell about our planet and the resources they provide for our future. Together, let us explore the fascinating world of sedimentary geology and its implications for our understanding of Earth's past, present, and future.

The Role of Technology in Sedimentary Geology

In recent years, advancements in technology have significantly enhanced the study of sedimentary rocks and the processes that shape them. This book acknowledges the importance of integrating modern tools and techniques into geological education and practice. Some of the key technological advancements include:

1. **Remote Sensing and GIS**: Geographic Information Systems (GIS) and remote sensing technologies allow geologists to analyze large datasets and visualize geological features from a distance. These tools are invaluable for mapping sedimentary basins, identifying potential resource deposits, and monitoring environmental changes over time.

2. **3D Modeling**: The use of three-dimensional modeling software enables geologists to create detailed representations of sedimentary structures and stratigraphy. This technology aids in visualizing complex geological formations and understanding the spatial relationships between different rock layers, which is crucial for resource exploration and environmental assessment.

3. **Geochemical Analysis**: Advances in geochemical analysis techniques, such as mass spectrometry and X-ray fluorescence, allow for precise characterization of sedimentary rock compositions. Understanding the chemical makeup of these rocks can provide insights into their formation processes and the environmental conditions at the time of deposition.

4. **Field Data Collection**: Mobile applications and digital tools facilitate real-time data collection in the field. Geologists can now record observations, take photographs, and analyze samples on-site, streamlining the research process and enhancing data accuracy.

5. **Educational Technology**: Online learning platforms and interactive simulations have transformed the way sedimentary geology is taught. These resources provide students with access to virtual field trips, interactive

exercises, and multimedia content that enrich the learning experience and make complex concepts more accessible.

Encouraging Sustainable Practices

As we explore the significance of sedimentary rocks, it is essential to emphasize the importance of sustainable practices in geology and resource management. The extraction and utilization of natural resources must be balanced with environmental stewardship to ensure the health of ecosystems and the well-being of communities.

1. **Responsible Mining**: For ASM practitioners, understanding the geological context of their operations is crucial for minimizing environmental impact. This book provides insights into sustainable mining practices, including techniques for reducing waste, managing water resources, and rehabilitating mined areas.

2. **Community Engagement**: Engaging local communities in the mining process fosters a sense of ownership and responsibility. By providing education and resources, ASM practitioners can empower communities to participate in sustainable practices that benefit both the environment and their livelihoods.

3. **Conservation of Geological Heritage**: Sedimentary rocks often contain valuable paleontological and archaeological records. Protecting these geological heritage sites is essential for preserving our understanding of Earth's history. This book encourages readers to appreciate the cultural and scientific significance of sedimentary formations and to advocate for their conservation.

Call to Action

As you engage with the content of this book, we encourage you to take an active role in the study and application of sedimentary geology. Whether you are a student, educator, or practitioner, consider the following actions:

- **Explore Local Geology**: Take the opportunity to study sedimentary rock formations in your area. Field trips and hands-on experiences can deepen your understanding and appreciation of geological processes.

- **Participate in Research**: Engage in research projects or internships that focus on sedimentary geology. Collaborating with experienced geologists can provide valuable insights and practical skills.
- **Advocate for Sustainability**: Promote sustainable practices within your community and industry. Share knowledge about responsible resource management and the importance of environmental stewardship.
- **Stay Informed**: Keep abreast of the latest developments in sedimentary geology and related fields. Attend conferences, workshops, and seminars to connect with other professionals and expand your knowledge base.

Final Reflections

In summary, "Sedimentary Rocks: A Comparative Illustrative Approach" is designed to be a comprehensive and engaging resource for anyone interested in the study of sedimentary geology. By combining theoretical knowledge with practical applications and modern technological advancements, this book aims to equip readers with the tools they need to navigate the complexities of the geological world.

As you journey through the pages, we hope you find inspiration in the stories that sedimentary rocks tell about our planet's past and the resources they hold for our future. Together, let us foster a deeper understanding of the Earth's processes and commit to responsible stewardship of its resources. Your exploration of sedimentary rocks is not just an academic pursuit; it is a vital step toward a sustainable future for our planet and its inhabitants.

The Interdisciplinary Nature of Sedimentary Geology

One of the remarkable aspects of sedimentary geology is its interdisciplinary nature. The study of sedimentary rocks intersects with various fields, enriching our understanding of Earth systems and enhancing practical applications. This book emphasizes the importance of integrating knowledge from different disciplines, including:

1. **Environmental Science**: Understanding sedimentary processes is crucial for addressing environmental issues such as soil erosion, sedimentation in waterways, and the impacts of climate change. By collaborating with environmental scientists, geologists can develop strategies to mitigate these challenges and promote sustainable land use.

2. **Paleontology**: Sedimentary rocks often contain fossils that provide insights into past life forms and ecosystems. The collaboration between geologists and paleontologists can lead to significant discoveries about evolutionary processes and the history of life on Earth. This book highlights the importance of fossils in interpreting sedimentary environments and understanding biological responses to environmental changes.

3. **Geochemistry**: The chemical composition of sedimentary rocks can reveal information about their origin and the conditions under which they formed. By integrating geochemical analysis with sedimentological studies, researchers can gain a more comprehensive understanding of sedimentary environments and the processes that influence them.

4. **Hydrology**: Sedimentary rocks play a critical role in groundwater systems. Understanding the porosity and permeability of these rocks is essential for managing water resources. Collaboration between geologists and hydrologists can lead to improved water management practices, particularly in regions reliant on groundwater for agriculture and drinking water.

5. **Engineering and Construction**: Knowledge of sedimentary rock properties is vital for construction and engineering projects. Understanding the geological context can inform decisions related to site selection, foundation design, and risk assessment for natural hazards such as landslides and flooding. This book provides insights into the practical applications of sedimentary geology in the construction industry.

The Global Context of Sedimentary Geology

Sedimentary rocks are found in diverse environments across the globe, each with unique characteristics and implications for resource management. This book encourages readers to consider the global context of sedimentary geology, including:

1. **Regional Variations**: Different regions exhibit distinct sedimentary rock types and formations, influenced by local geological history, climate, and tectonic activity. Understanding these regional variations is essential for effective resource exploration and environmental management.

2. **Global Challenges**: Issues such as climate change, resource depletion, and habitat destruction are global challenges that require a comprehensive

understanding of sedimentary processes. By studying sedimentary rocks, we can gain insights into past climate conditions and better predict future changes, informing strategies for adaptation and mitigation.

3. **Cultural Significance**: Sedimentary rocks often hold cultural and historical significance for local communities. Recognizing and respecting the cultural heritage associated with geological formations is essential for fostering sustainable practices and promoting community engagement in resource management.

Future Research Directions

As the field of sedimentary geology continues to evolve, several areas of research hold promise for advancing our understanding of sedimentary processes and their implications:

1. **Climate Change Studies**: Investigating sedimentary records can provide valuable insights into past climate conditions and help predict future climate scenarios. Research in this area can inform strategies for climate adaptation and resilience.

2. **Resource Sustainability**: Ongoing research into sustainable extraction methods and resource management practices is crucial for ensuring that sedimentary resources are utilized responsibly. This includes exploring alternative materials and recycling practices to reduce reliance on non-renewable resources.

3. **Technological Innovations**: Continued advancements in technology, such as artificial intelligence and machine learning, have the potential to revolutionize sedimentary geology. These tools can enhance data analysis, improve predictive modeling, and facilitate more efficient resource exploration.

4. **Public Engagement and Education**: Increasing public awareness and understanding of sedimentary geology is essential for fostering a culture of sustainability. Research into effective educational strategies and outreach programs can help bridge the gap between scientific knowledge and public understanding.

Closing Remarks

In conclusion, "Sedimentary Rocks: A Comparative Illustrative Approach" is designed to be a comprehensive and engaging resource that not only educates but also inspires action. By exploring the intricate relationships between sedimentary rocks, Earth processes, and human activities, we hope to empower readers to become informed stewards of our planet.

As you navigate through the chapters, we encourage you to reflect on the broader implications of sedimentary geology in addressing contemporary challenges. Whether you are a student, educator, or practitioner, your engagement with this field can contribute to a more sustainable future.

We invite you to embrace the journey of discovery that lies ahead. The world of sedimentary rocks is rich with stories waiting to be uncovered, and your exploration can lead to new insights and innovations that benefit both society and the environment. Together, let us deepen our understanding of the Earth's processes and work towards a sustainable future for generations to come.

Engaging with the Community: The Role of Outreach and Education

As we conclude our exploration of sedimentary rocks and their significance, it is essential to highlight the importance of community engagement and education in fostering a deeper understanding of geology. Outreach initiatives can play a pivotal role in bridging the gap between scientific knowledge and public awareness. Here are several strategies to enhance community engagement:

1. **Workshops and Seminars**: Organizing workshops and seminars for local communities can provide valuable opportunities for hands-on learning. These events can cover topics such as sedimentary rock identification, resource management, and sustainable practices. Engaging local experts and educators can enhance the effectiveness of these initiatives.

2. **Field Trips and Excursions**: Field trips to local geological sites can offer participants a firsthand experience of sedimentary formations and their significance. Guided tours led by knowledgeable geologists can help demystify geological processes and foster a sense of connection to the Earth.

3. **Collaborative Projects**: Partnering with schools, universities, and community organizations to develop collaborative projects can enhance educational outreach. These projects can focus on local geological studies,

environmental conservation efforts, or resource management initiatives, encouraging community involvement and ownership.

4. **Public Lectures and Events**: Hosting public lectures and events featuring guest speakers from the field of geology can stimulate interest and curiosity. These events can cover a range of topics, from the basics of sedimentary geology to current research and its implications for society.

5. **Utilizing Digital Platforms**: In the digital age, leveraging online platforms for education and outreach is crucial. Creating engaging content, such as videos, podcasts, and interactive webinars, can reach a broader audience and make geological concepts more accessible. Social media can also be a powerful tool for sharing knowledge and fostering discussions about sedimentary geology.

The Importance of Lifelong Learning

The study of sedimentary rocks and geology is not confined to formal education; it is a lifelong journey of discovery. As new research emerges and technologies evolve, staying informed and engaged is essential for anyone interested in this field. Here are some ways to promote lifelong learning:

1. **Continuing Education Programs**: Many institutions offer continuing education courses and certifications in geology and related fields. These programs can help professionals stay updated on the latest developments and enhance their skills.

2. **Professional Associations**: Joining professional organizations, such as geological societies or mining associations, can provide access to resources, networking opportunities, and professional development. These organizations often host conferences, workshops, and publications that keep members informed about advancements in the field.

3. **Research Participation**: Engaging in research projects, whether independently or collaboratively, can deepen understanding and contribute to the body of knowledge in sedimentary geology. Participating in citizen science initiatives can also provide valuable data for ongoing research.

4. **Reading and Self-Study**: Encouraging a habit of reading scientific literature, textbooks, and popular science articles can foster a deeper understanding of sedimentary geology. Online resources, such as academic

journals and educational websites, can provide valuable insights into current research and trends.

Final Thoughts on the Journey Ahead

As we conclude this comprehensive exploration of sedimentary rocks and their significance, we hope that *"Sedimentary Rocks: A Comparative Illustrative Approach"* serves as a valuable resource for your academic and professional journey. The knowledge and insights contained within these pages are intended to inspire curiosity, critical thinking, and a commitment to responsible stewardship of our planet.

The study of sedimentary rocks transcends mere academic pursuit; it is a vital endeavor that connects us to the Earth's history, informs our understanding of current environmental challenges, and guides our efforts toward a sustainable future. Sedimentary rocks, formed through processes of weathering, erosion, and deposition, hold within them the narratives of ancient environments, climate changes, and biological evolution. By studying these rocks, we gain insights not only into the geological past but also into the present-day processes that shape our landscapes and ecosystems.

As you engage with the material, we encourage you to reflect on the broader implications of your learning. Consider how the principles of sedimentary geology can inform your understanding of pressing issues such as climate change, resource management, and environmental conservation. The knowledge you acquire can empower you to contribute meaningfully to the field of geology and to the well-being of our planet. Whether through research, education, or community engagement, your efforts can help foster a deeper appreciation for the Earth's resources and the importance of sustainable practices.

Moreover, the interdisciplinary nature of geology allows for collaboration with other fields such as environmental science, archaeology, and engineering. This interconnectedness opens up numerous avenues for exploration and innovation. As you delve deeper into sedimentary geology, consider how your findings can intersect with other disciplines to address complex global challenges. For instance, understanding sedimentary processes can enhance our ability to manage water resources, mitigate natural hazards, and restore ecosystems.

In closing, we invite you to embrace the wonders of sedimentary geology. Explore the diverse landscapes around you, engage with your community, and continue to seek knowledge and understanding. The stories that sedimentary rocks tell are waiting to be uncovered, and your journey into this fascinating field has the potential to make a meaningful impact on both your life and the world around you.

As you embark on this journey, remember that geology is not just about rocks; it is about understanding the intricate relationships between the Earth's systems and the life they support. Each layer of sediment tells a story, and each fossil embedded within it offers a glimpse into a world long gone. By studying these layers, you become a storyteller of the Earth, interpreting its past and informing its future.

Together, let us celebrate the dynamic processes that shape our planet and work towards a sustainable future for generations to come. Your passion and dedication can inspire others, and together, we can cultivate a community of informed stewards committed to protecting our planet's geological heritage. As you move forward, may you carry with you the spirit of inquiry and the desire to make a positive impact, not only in your own life but also in the lives of those around you and the health of our planet. The journey ahead is filled with opportunities for discovery, and we are excited to see how you will contribute to the ever-evolving narrative of Earth's geology.

FSC
www.fsc.org
MIX
Papier | Fördert
gute Waldnutzung
FSC® C083411

Kolibri 360 GmbH
Ferdinand-Jühlke-Straße 7
99095 Erfurt, Deutschland
produktsicherheit@kolibri360.de

Druck:
CPI Druckdienstleistungen GmbH
im Auftrag der
Kolibri 360 GmbH
Ferdinand-Jühlke-Str. 7
99095 Erfurt